Promoting Global Environmental Sustainability and Cooperation

Sofia Idris
Independent Researcher, Pakistan

A volume in the Practice, Progress, and Proficiency in Sustainability (PPPS) Book Series

Published in the United States of America by
IGI Global
Engineering Science Reference (an imprint of IGI Global)
701 E. Chocolate Avenue
Hershey PA, USA 17033
Tel: 717-533-8845
Fax: 717-533-8661
E-mail: cust@igi-global.com
Web site: http://www.igi-global.com

Copyright © 2018 by IGI Global. All rights reserved. No part of this publication may be reproduced, stored or distributed in any form or by any means, electronic or mechanical, including photocopying, without written permission from the publisher.
Product or company names used in this set are for identification purposes only. Inclusion of the names of the products or companies does not indicate a claim of ownership by IGI Global of the trademark or registered trademark.

Library of Congress Cataloging-in-Publication Data

Names: Idris, Sofia, 1986- editor.
Title: Promoting global environmental sustainability and cooperation / Sofia
   Idris, editor.
Description: Hershey, PA : Engineering Science Reference, [2018]
Identifiers: LCCN 2017025742| ISBN 9781522539902 (hardcover) | ISBN
   9781522539919 (ebook)
Subjects: LCSH: Sustainable development. | Globalization. | International
   cooperation. | Environmental policy.
Classification: LCC HC79.E5 P7265 2018 | DDC 338.9/27--dc23 LC record available at https://lccn.loc.gov/2017025742

This book is published in the IGI Global book series Practice, Progress, and Proficiency in Sustainability (PPPS) (ISSN: 2330-3271; eISSN: 2330-328X)

British Cataloguing in Publication Data
A Cataloguing in Publication record for this book is available from the British Library.

All work contributed to this book is new, previously-unpublished material.
The views expressed in this book are those of the authors, but not necessarily of the publisher.

For electronic access to this publication, please contact: eresources@igi-global.com.

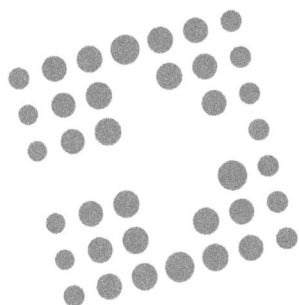

Practice, Progress, and Proficiency in Sustainability (PPPS) Book Series

ISSN:2330-3271
EISSN:2330-328X

Editor-in-Chief: Ayman Batisha, International Sustainability Institute, Egypt

MISSION

In a world where traditional business practices are reconsidered and economic activity is performed in a global context, new areas of economic developments are recognized as the key enablers of wealth and income production. This knowledge of information technologies provides infrastructures, systems, and services towards sustainable development.

The **Practices, Progress, and Proficiency in Sustainability (PPPS) Book Series** focuses on the local and global challenges, business opportunities, and societal needs surrounding international collaboration and sustainable development of technology. This series brings together academics, researchers, entrepreneurs, policy makers and government officers aiming to contribute to the progress and proficiency in sustainability.

COVERAGE

- Socio-Economic
- Intellectual Capital
- E-Development
- Knowledge clusters
- Innovation Networks
- Eco-Innovation
- Strategic Management of IT
- ICT and knowledge for development
- Technological learning
- Global Business

IGI Global is currently accepting manuscripts for publication within this series. To submit a proposal for a volume in this series, please contact our Acquisition Editors at Acquisitions@igi-global.com or visit: http://www.igi-global.com/publish/.

The Practice, Progress, and Proficiency in Sustainability (PPPS) Book Series (ISSN 2330-3271) is published by IGI Global, 701 E. Chocolate Avenue, Hershey, PA 17033-1240, USA, www.igi-global.com. This series is composed of titles available for purchase individually; each title is edited to be contextually exclusive from any other title within the series. For pricing and ordering information please visit http://www.igi-global.com/book-series/practice-progress-proficiency-sustainability/73810. Postmaster: Send all address changes to above address. ©© 2018 IGI Global. All rights, including translation in other languages reserved by the publisher. No part of this series may be reproduced or used in any form or by any means – graphics, electronic, or mechanical, including photocopying, recording, taping, or information and retrieval systems – without written permission from the publisher, except for non commercial, educational use, including classroom teaching purposes. The views expressed in this series are those of the authors, but not necessarily of IGI Global.

Titles in this Series

For a list of additional titles in this series, please visit:
htttps://www.igi-global.com/book-series/practice-progress-proficiency-sustainability/73810

Driving Green Consumerism Through Strategic Sustainability Marketing
Farzana Quoquab (Universiti Teknologi Malaysia, Malaysia) Ramayah Thurasamy (Universiti Sains Malaysia, Malaysia) and Jihad Mohammad (Universiti Teknologi Malaysia, Maaysia)
Business Science Reference • ©2018 • 301pp • H/C (ISBN: 9781522529125) • US $215.00

Entrepreneurship and Business Development in the Renewable Energy Sector
Adrian Dumitru Tantau (Bucharest University of Economic Studies, Romania) and Laurențiu Cătălin Frățilă (Bucharest University of Economic Studies, Rmania)
Business Science Reference • ©2018 • 381pp • H/C (ISBN: 9781522536253) • US $245.00

Technological Tools for Value-Based Sustainable Relationships in Health Emerging...
Mosad Zineldin (Linnaeus University, Sweden) and Valentina Vasicheva (Linnaeus University, Sweden)
Medical Information Science Reference • ©2018 • 149pp • H/C (ISBN: 9781522540915) • US $155.00

Sustainable Local Energy Planning and Decision Making Emerging...
Vangelis Marinakis (National Technical University of Athens, Greece)
Information Science Reference • ©2017 • 171pp • H/C (ISBN: 9781522522867) • US $140.00

Sustainable Potato Production and the Impact of Climate Change
Sunil Londhe (International Centre for Research in Agroforestry (ICRAF), India)
Information Science Reference • ©2017 • 323pp • H/C (ISBN: 9781522517153) • US $180.00

Handbook of Research on Green Economic Development Initiatives and Strategies
M. Mustafa Erdoğdu (Marmara University, Turkey) Thankom Arun (University of Essex, UK) and Imran Habib Ahmad (Global Green Growth Institute, South Korea)
Business Science Reference • ©2016 • 747pp • H/C (ISBN: 9781522504405) • US $335.00

Impact of Water Pollution on Human Health and Environmental Sustainability
A. Elaine McKeown (Independent Researcher, USA) and George Bugyi (Pennsylvania State University, USA)
Information Science Reference • ©2016 • 422pp • H/C (ISBN: 9781466695597) • US $220.00

For an entire list of titles in this series, please visit:
https://www.igi-global.com/book-series/practice-progress-proficiency-sustainability/73810

701 East Chocolate Avenue, Hershey, PA 17033, USA
Tel: 717-533-8845 x100 • Fax: 717-533-8661
E-Mail: cust@igi-global.com • www.igi-global.com

Table of Contents

Preface xii

Acknowledgment xx

Introduction xxi

Section 1
Globalization and Global Governnce

Chapter 1
Globalization and Rethinking of Environmental Consumption From a
Sustainability Perspective 1
    Luke A. Amadi, University of Port Harcourt, Nigeria
    Prince Ikechukwu Igwe, University of Port Harcourt, Nigeria

Chapter 2
From Kyoto to Paris: An Analysis of the Politics of Multilateralism on
Climate Change 31
    Moses Metumara Duruji, Covenant University, Nigeria
    Faith O. Olanrewaju, Covenant University, Nigeria
    Favour U. Duruji-Moses, Covenant University, Nigeria

Section 2
Regional Politics

Chapter 3
Water Security in Pakistan 58
    Sofia Idris, Independent Researcher, Pakistan

Section 3
International Business and Consumer Behavior

Chapter 4
International Businesses and Environmental Issues 86
Fatma Ince, Mersin University, Turkey

Chapter 5
Consumer Cooperation in Sustainability: The Green Gap in an Emerging
Market 112
Njabulo Mkhize, University of kwaZulu-Natal, South Africa
Debbie Ellis, University of KwaZulu-Natal, South Africa

Section 4
Sustainable Development

Chapter 6
"Airpocalypse" or Tsar Economic China: Analysis of Unsustainable
Environment and Reason Behind Increased Global Warming..........................137
Fauzia Ghani, G. C. University, Pakistan
Komal Ashraf Qureshi, Kinnaird College for Women, Pakistan

Chapter 7
Global Implications of Sustainability and E-Society Infrastructure in
Developing Economies 162
Birău Ramona, Constantin Brâncusi University, Romania

Related References 184

Compilation of References 226

About the Contributors 255

Index 258

Detailed Table of Contents

Preface xii

Acknowledgment xx

Introduction xxi

Section 1
Globalization and Global Governnce

Chapter 1
Globalization and Rethinking of Environmental Consumption From a
Sustainability Perspective 1
    Luke A. Amadi, University of Port Harcourt, Nigeria
    Prince Ikechukwu Igwe, University of Port Harcourt, Nigeria

Sustainable environmental consumption has been a marginalized concept in international development studies and cooperation. In recent decades, there has been growing interest in identifying robust indicators that demonstrate the evidence of globalization and unsustainable environmental consumption. Globalization is premised on integrating the world into a global village. Various dimensions of globalization have different effects on the ecosystem. Plausible evidence linking globalization trajectories into practical interactions suggesting sustainable environmental consumption has been less lucid as the effects of globalization on the ecological environment does not provide clear patterns. This hugely significant problem has reopened critical debates on novel thinking on dynamics of environmental consumption patterns of the affluent societies in the era of globalization and its implications on environmental sustainability. This chapter deployed content analysis methodology and political ecology framework to review and analyze seminal studies on sustainable consumption and globalization, including relevant globalization indexes. The aim is to provide evidence of the impact of globalization on environmental consumption. The chapter suggests that globalization results in asymmetrical and deleterious natural resource extraction between the affluent North and poor South. It offered alternative thinking in which sustained policy framings and international development collaboration could be institutionalized to strengthen sustainable environmental consumption one which is premised on ecological justice and natural resource equality.

Chapter 2
From Kyoto to Paris: An Analysis of the Politics of Multilateralism on
Climate Change 31
    Moses Metumara Duruji, Covenant University, Nigeria
    Faith O. Olanrewaju, Covenant University, Nigeria
    Favour U. Duruji-Moses, Covenant University, Nigeria

The Earth Summit of 1992 held in Rio de Janeiro awakened the consciousness of the world to the danger of climate change. The establishment of the United Nations Framework Convention on Climate Change provided the platform for parties to negotiate on ways of moving forward. The global acknowledgement of the weightiness of the climate change and the future of the planet galvanized international agreements to this regard. Consequently, a landmark agreement was brokered in 1992 at Kyoto, Japan and 2015 in Paris, France. However, the strong issues of national interest tend to bedevil the implementation that would take the world forward on climate change. The chapter therefore examined multilateralism from the platform of climate change conferences and analyzed the political undertone behind disappointing outcomes even when most of the negotiators realized that the only way to salvage the impending doom is a multilateral binding agreement when nation-state can subsume their narrow interest.

Section 2
Regional Politics

Chapter 3
Water Security in Pakistan 58
    Sofia Idris, Independent Researcher, Pakistan

Pakistan largely faces water scarcity, and the arid agricultural land of Pakistan is mainly due to the violation of the Indus Water Treaty by its neighbor India. By blocking the river flow towards Pakistan from the head-works, India has been building excessive dams, barrages, and power projects which are illegal according to the above-mentioned international treaty, and Pakistan has many times appealed in the UN to take action against this unfair act. However, so far, nothing could be done in this regard. The study will be helpful to understand the various challenges facing Pakistan to cater the insufficient supply of water and will give insight on the most important dimensions and facts about the international challenges to meet the shortage. The trans-boundary water issue between China and India have also been studied to try to explore new options and find the solution of a much pressing problem. The study might thus contribute to understand the issue, study the role of international community, and give useful and practical suggestions to solve the most pressing problem.

Section 3
International Business and Consumer Behavior

Chapter 4
International Businesses and Environmental Issues 86
    Fatma Ince, Mersin University, Turkey

This chapter on international businesses and environmental issues addresses the relationship between activities of international businesses and environmental goals. Because of the increasing awareness about environmental issues, the related groups, international policies, and relation force the businesses to be more environmentally friendly and consider the future of resources in their operations. The international regulations, declarations, and other pressures can change the usages of the businesses. The combination of the planet, profit, and people are conceived to protect nature with the aim of growth. From this viewpoint, this chapter provides an overview of the environmental issues, international politics, relations, standards, and successful examples of businesses about being environmentally conscious.

Chapter 5
Consumer Cooperation in Sustainability: The Green Gap in an Emerging Market 112
    Njabulo Mkhize, University of kwaZulu-Natal, South Africa
    Debbie Ellis, University of KwaZulu-Natal, South Africa

The planet is under threat. Unless all stakeholders, that is governments, businesses, and consumers, become more environmentally friendly, some predict dire consequences for the earth and all those who inhabit it. While governments and businesses have a role to play, green consumer behavior is vital to the sustainability of the environment. Consumers have been shown to express increasing concern for the environment, but in many studies this concern has been found not to be matched by actions, a phenomenon labelled the green gap. This chapter describes a study that investigated the existence and extent of the green gap amongst a sample of South African adult consumers and sought to determine possible reasons for a lack of green behavior. Recommendations are made to marketers and policy makers to encourage consumer cooperation in environmental sustainability.

Section 4
Sustainable Development

Chapter 6
"Airpocalypse" or Tsar Economic China: Analysis of Unsustainable
Environment and Reason Behind Increased Global Warming..........................137
    Fauzia Ghani, G. C. University, Pakistan
    Komal Ashraf Qureshi, Kinnaird College for Women, Pakistan

This chapter focuses on the case study of China, which is facing grave issues regarding environment and global warming. Hence, the "Airpocalypse" in China led to need and debate about the sustainability of the environment. In this chapter, an effort has been made to analyze the environmental sustainability risk which the country of China can have for the increasing rate of global warming, and how this part of region can have a transnational impact on other neighboring countries when it comes to the cause of making environment pure from pollutants, carbon dioxide, and coal emissions. The methodology of this research is qualitative, descriptive, and analytical in nature. This chapter includes the variable of environmental sustainability which is dependent on the energy consumption of industries of China involved in emission of greenhouse gases.

Chapter 7
Global Implications of Sustainability and E-Society Infrastructure in
Developing Economies 162
    Bîrău Ramona, Constantin Brâncusi University, Romania

The main purpose of this chapter is to investigate the global implications of sustainability and e-society infrastructure in developing economies. A generally accepted definition of e-society is very difficult to compress into words considering the complexity of the phenomenon itself. However, an exhaustive approach includes a great variety of original views and individual multifaceted opinions which converge to obtain a solid theoretical structure. Globalization is the modern term used to describe changes in the structure of societies and the world economy, but having a major impact in the context of an accelerated informatization. The process of globalization is not a new and innovative process, but it is the result of changes in the world economy that have increased in recent years, considering the fact that it brings a number of advantages. Moreover, globalization means labor mobility without constraint of geographical boundaries. Generally, the progress of communications is another consequence of globalization and the impact of change is even stronger in developing economies.

Related References 184

Compilation of References 226

About the Contributors 255

Index 258

Preface

ENVIRONMENT

According to the Oxford Advanced Learner's Dictionary, environment is: "The natural world in which people, animals and plants live" (Wehmier & Ashby, 2000).

Merriam-Webster defines environment as "the circumstances, objects, or conditions by which one is surrounded." And: "the complex of physical, chemical, and biotic factors (such as climate, soil, and living things) that act upon an organism or an ecological community and ultimately determine its form and survival" (Environment, n.d.).

English Oxford Living Dictionaries defines environment as "The surroundings or conditions in which a person, animal, or plant lives or operates." Also: "The natural world, as a whole or in a particular geographical area, especially as affected by human activity" (Environment, n.d.).

ENVIRONMENTAL SUSTAINABILITY

The concept of sustainable development was initially introduced in 1972. It was adopted as a vision recognizing the interconnectedness of environmental, social, and economic issues (Asian Development Bank [ADB], 2012).

However, Thomas Malthus had showed his concern about environmental sustainability in his Essay on the Principle of Population in 1798. In his seminal work, Malthus stated the potential for geometric increase in population. "The power of population is indefinitely greater than the power of the earth to produce subsistence for man" (Malthus, 1998; Santen, 2013; World Economic Forum [WEF], 2013).

John Morelli defines environmental sustainability as "a condition of balance, resilience, and interconnectedness that allows human society to satisfy its needs while neither exceeding the capacity of its supporting ecosystems to continue to regenerate the services necessary to meet those needs nor by our actions diminishing biological diversity" (Morelli, 2011).

Preface

WHY ENVIRONMENT SHOULD BE PROTECTED?

Nature has been changing constantly since the very beginning. There are many factors that are responsible for this change. These factors include both the natural causes as well as the ones created by human activity. From the Sahara Pump to the Ice Age, from the breakup of Pangaea to the extinction of dinosaurs, Earth has entirely changed from that which was millions of years ago. Man has had careless interaction with the surrounding environment by continuously adapting the Plane in order to accommodate their needs.

In the words of French philosopher Rene Descartes, "the main theme of modern science is to equip humans with the tool to conquer over matter and become lords and master of nature" (Nazaruk, 2016).

So why should we protect our environment? Aren't we supposed to extract goods and resources from nature for our sustenance, subsistence and other requirements? Is it or should it be a give and take relationship between mankind and the nature? Should we care enough to restore and preserve the nature?

Modern science has provided huge empirical evidence which revealed that nature was a complex collection of air, water, soil, plants, animal and human beings. The components were interdependent enough that the breakdown of one part of the system could damage the productivity of other parts. Nature is not merely a warehouse of resources to serve human needs. Instead, it is interdependent, a highly integrated functioning system upon which all life forms, including the plants, animals, water, soil, and humans depend for survival. Therefore, it is necessary for mankind to take care of the natural environment with love and respect for the reason that the failure of the system would eventually threaten the subsistence of human beings (Nazaruk, 2016).

The environment affects us as much as we effect our environment. For instance, deforestation for housing and other development projects doesn't only harm the nature; it also contributes to air pollution and global warming too. The chemical and other waste dumping into water such as oil spillage, doesn't only affect marine life, it also has the potential to disturb the food chain as well as creating a shortage of clean drinking water. Similarly, massive cutting of trees may lead to soil erosion and salinity in the soil as a result; the once fertile land may become arid agriculture land and; ultimately, in extreme cases, may even lead to desertification. The once fertile land will now be unfit for cultivation resulting in cases ranging from low cases of food shortage to extreme cases of famine. Take air for another example, air pollution includes particulates, nitrogen oxides, sulfur dioxide, and in many cases, ozone epidemiology studies are impacted by more than one pollutant. It is established that these pollutants are fundamentally related to cardiovascular disease (Erickson, Griswold, Maghirang, & Urbas-Zewski, 2017).

These pollutants in the air may have played a part in individuals with asthma and/or chronic obstructive pulmonary disease they breathe. Lung cancer mortality is related with particulates from diesel engines. Evidence shows that compounds in metropolitan air pollutants reach the blood stream and the brain through the lungs. Decreased cognitive function, Parkinson's disease, dementia and autism in children are identified as being more widespread in large cities owing to air pollution. Traffic related studies have revealed that dementia is linked with air pollution from vehicles with small particulates. The correlation of Parkinson's disease with air pollution is found to be not as strong as with dementia. About 47 Million people around the world are living with dementia, and the projected annual cost is $818 billion. There are a number of studies that show an increase in the percentage of children with autism as a result of increased concentrations of air pollutants. The associations were stronger for boys as compared to the girls (Erickson, Griswold, Maghirang, & Urbas-Zewski, 2017).

NEED FOR ENVIRONMENTAL SUSTAINABILITY AND GLOBAL COOPERATION

In 1987, a document entitled "Our Common Future," commonly known as the "Brundtland Report" was prepared by the World Commission for Environment and Development (WCED) with the coordination of commission, Gro Harlem Brundtland; giving the concept of sustainable development.

Sustainable development was thus defined as "sustainable development is the development that meets the needs of the present without compromising the ability of future generations to meet their own needs" (Vezzoli & Manzini, 2008).

This definition refers to the well-being of people as an environmental quality although it does not directly refer to the environment. This originates a fundamental ethical principle of "responsibility of the present generations to future generations" (Vezzolic& Manzini, 2008).

Brundtland Report is historically important for environmental sustainability for two reasons. First, it opened up international debate on aforementioned definitive imposition of the idea of our responsibility towards the future. Secondly, it questioned the ideas of development, which was previously indisputable. Subsequently, several other important and prominent international initiatives were started. The significant initiatives among these include: UN Conference on Environment and Development in 1992 was held in Rio de Janeiro and World Summit on Sustainable Development that was held ten years later in Johannesburg (Vezzoli & Manzini, 2008). Subsequently, UN Conference on Sustainable Development (Rio+20) was held in 2012. The Sustainable Development Goals (SDGs) also known as the Global Goals, were

Preface

developed at the United Nations Conference on Sustainable Development in Rio de Janeiro in 2012. The SDGs replaced the Millennium Development Goals (MDGs). The Millennium Development Goals (MDGs) had previously started a global effort in 2000 to tackle a total of 08 Goals which included the goal to ensure environmental sustainability (World Health Organization [WHO], n.d.).

Globalization has integrated the world nations to cooperate with each other in order to promote peace and prosperity and peacefully resolving problems affecting them. Thus, the world became a global village where the countries are now determined to cooperate with each other realize global agendas and help implement those policies to achieve such common goals. These agendas or goals include Environmental Sustainability.

THE CHALLENGES

Globalization has brought the world nations together promising to end discrimination and deprivation of the developing countries and progress together by strengthening international cooperation and therefore, the international relations. Hence, the world became a global village whereby the states are now resolute to cooperate with each other realize global agendas and help implement those policies to achieve such common goals. These agendas or goals include Environmental Sustainability such as Millennium Development Goals (MDGs) and Sustainable Development Goals (SDGs) among other initiatives.

Earlier, international relations didn't focus on environmental issues and were generally considered a state's internal issue that required domestic technical solutions rather than multilateral political solutions. States didn't give such issues a priority as part of their foreign policy. The beginning of the twenty-first century witnessed developments towards raising awareness on environmental issues and of their relationship with development, under the notion of "sustainability", and in establishing a multilateral framework including the state and non-state actors to realize this goal of sustainability.

The development of the book has been a process that required a lot of hard work and diligence. The authors of this book come from all parts of the globe and are the scholars from diverse backgrounds. This book has taken advantage from their expertise in diverse subject areas of the theme of this book. The authors had been cooperative and diligent. They had been very enthusiastic towards this book project. As mentioned earlier, the authors belong to different regions of the world, including the region where power was available for only three hours a day. At times, there were some technical difficulties with the authors; but this didn't stop them from believing in this project. Their cooperation, diligence and hard work are the reason that we have a book project brought to its completion.

Being the first ever editing a full book project, it was an adventure and a difficult journey for the editor as well. Without any other editors in the project, I had to be very careful and double check everything before proceeding to the next phase of the publishing process; thus keeping a slow pace. I just had to be confident that everything was precisely done as it was supposed to be.

SEARCHING FOR A SOLUTION

It is very hard to find literature on environmental sustainability. There may be a lot of literature available regarding the sustainable development that involved development in many areas but there was a lack of scholarly research on environmental issues. One can find studies on sustainable development that would cover issues based on different matters such as economics, education, logistics, industries and even tourism etc. but it hardly covers environmental issues.

Moreover, there is a lack of scholarship on the global efforts to promote environmental sustainability and reduce environmental degradation. Although there is a global consciousness on the importance of environmental sustainability and there has been much concentration on working on projects that promote environmental sustainability. The global community now interacts more closely and more diligently work together thanks to globalization. It is now more open and more cooperative to help each other to promote peace and sustainability. But competition can also be observed with this cooperation. Among other factors, global financial crisis brought with it a stricter economic competition between the world nations thus, foregoing the issue of environmental sustainability. The need to promote environmental sustainability in the world requires global cooperation. Nevertheless, the scholarship on such issues and efforts at the international level is also difficult to find.

This book includes some case studies of various countries as well as research on patterns of consumption, international business and world economies. Furthermore, this book looks to discuss and address the difficulties and challenges of achieving environmental sustainability. The chapters in this book address different aspects of international relations, ranging from foreign policies, cooperation and competition, conflict and conflict resolution, trade, business, technology, economics and politics and related topics concerning environmental sustainability. Additionally, the book will explore the changing world situation in the contemporary globalized world with some states exerting their influence subsiding the spirit of cooperation and overlooking the issue of environmental sustainability.

Preface

ORGANIZATION OF THE BOOK

The book is organized into seven chapters. A brief description of each of the chapters follows:

Introduction

The book opens with the introduction of this book. The first chapter therefore, offers an understanding about the concepts used and the theme of the book. It also gives account of what this book is all about.

Chapter 1

Chapter 1 reviewed and analyzed seminal studies on sustainable consumption and globalization to provide evidence of the impact of globalization on environmental consumption.

Chapter 2

This chapter examines multilateralism from the platform of climate change conferences and analyzed the political undertone behind disappointing outcomes even when most of the negotiators realized that the only way to salvage the impending doom is a multilateral binding agreement when nation-state can subsume their narrow interest.

Chapter 3

Analyzes the various challenges facing Pakistan to cater the insufficient supply of water and will give insight on the most important dimensions and facts about the international challenges to meet the shortage. A similar case study on China has also been included in an attempt to learn new options and find the solution to this latent problem.

Chapter 4

It is an overview of the environmental issues, international politics, relations, standards and successful examples of businesses about being environmentally conscious.

Chapter 5

It is a study which investigated the existence and extent of the green gap among South African adult consumers. It also studies possible reasons for a lack of green behaviour. The chapter also gives recommendations to the marketers and policy makers to promote consumer cooperation in environmental sustainability.

Chapter 6

Analyze the environmental sustainability risk which China can have for the increasing rate of global warming; and how this part of region can have a transnational impact on other neighboring countries when it comes to the cause of making environment pure from pollutants, carbon dioxide, and coal emissions.

Chapter 7

The book ends with the chapter which investigates the global implications of sustainability and e-society infrastructure in developing economies.

REFERENCES

Asian Development Bank (ADB). (2012). *World Sustainable Development Timeline*. Retrieved from https://www.researchgate.net/profile/Olivier_Serrat/publication/266878643_World_Sustainable_Development_Timeline/links/543e3e430cf2d6934ebd20af/World-Sustainable-Development-Timeline.pdf?origin=publication_list

Environment. (n.d.a). In *English Oxford Living Dictionaries*. Retrieved from: https://en.oxforddictionaries.com/definition/environment

Environment. (n.d.b). In *Merriam-Webster Dictionary*. Retrieved from: https://www.merriam-webster.com/dictionary/environment

Erickson, L. E., Griswold, W., Maghirang, R. G., & Urbas-zewski, B. P. (2017). Air Quality, Health and Community Action. *Journal of Environmental Protection*, 8(10), 1057–1074. doi:10.4236/jep.2017.810067

Malthus, T. (1998). *An Essay on the Principle of Population*. London: Electronic Scholarly Publishing Project. Available at: http://129.237.201.53/books/malthus/population/malthus.pdf

Morelli, J. (2011). Environmental Sustainability: A Definition for Environmental Professionals. *Journal of Environmental Sustainability*, *1*(1), 1–10. doi:10.14448/jes.01.0002

Nazaruk, P. A. (2016). Why Should We Take Care Of Nature? *HuffPost*. Retrieved from: https://www.huffingtonpost.com/pawel-alva-nazaruk/why-should-we-take-care-o_b_12170852.html

"Science for Environment Policy": European Commission DG Environment News Alert Service. (2016, May 6). SCU, The University of the West of England.

van Santen, L. (2013). *Timeline: a brief history of sustainable consumption*. World Economic Forum. Retrieved from: https://www.weforum.org/agenda/2013/11/timeline-a-brief-history-of-sustainable-consumption/

Vezzoli, C. A., & Manzini, E. (2008). *Design for Environmental Sustainability*. Milan: Springer Science & Business Media.

Wehmier, S., & Ashby, M. (Eds.). (2000). *Oxford Advanced Learner's Dictionary*. Oxford, UK: Oxford University Press.

World Economic Forum (WEF). (2013). *Sustainable Consumption: Stakeholder Perspectives*. Retrieved from: http://www3.weforum.org/docs/WEF_ENV_SustainableConsumption_Book_2013.pdf

World Health Organization (WHO). (n.d.). *Millennium Development Goals (MDGs)*. Retrieved from: http://www.who.int/topics/millennium_development_goals/about/en/

Acknowledgment

The editor would like to acknowledge the help of all the people involved in this project and, more specifically, to the authors and reviewers that took part in the review process. Without their support, this book would not have become a reality.

First, the editor would like to thank each one of the authors for their contributions. My sincere gratitude goes to the chapter's authors who contributed their time and expertise to this book.

Second, the editor wishes to acknowledge the valuable contributions of the reviewers regarding the improvement of quality, coherence, and content presentation of chapters. Furthermore, the valuable contribution of Dr. Ayşenur Erdil, Marmara University, Turkey; Dr. Kadir Dede, Hacettepe University, Turkey; Dr. Kijpokin Kasemsap, Suan Sunandha Rajabhat University, Thailand and Prof. Pedro Pinheiro Gomes, National Statistics Institute of Portugal, Portugal in the double-blind peer review process is highly acknowledged.

Sofia Idris
Independent Researcher, Pakistan

Introduction

ENVIRONMENTAL SUSTAINABILITY

Environmental sustainability may be defined as "a condition of balance, resilience, and interconnectedness that allows human society to satisfy its needs while neither exceeding the capacity of its supporting ecosystems to continue to regenerate the services necessary to meet those needs nor by our actions diminishing biological diversity" (Morelli, 2011).

The concept of sustainable development was initially introduced in 1972. It was adopted as a vision recognizing the interconnectedness of environmental, social, and economic issues (Asian Development Bank [ADB], 2012).

The Sustainable Development Goals (SDGs) also known as the Global Goals, were developed at the United Nations Conference on Sustainable Development in Rio de Janeiro in 2012. The purpose was to produce a set of universal goals that meet the urgent political, economic and environmental challenges facing our world (United Nations Development Programme [UNDP], n.d.).

There are 17 Sustainable Development Goals (SDGs). These are: no poverty, zero hunger, good health and wellbeing, quality education, gender equality, clean water and sanitation, affordable and clean energy, decent work and economic growth, industry innovation and infrastructure, reduced inequalities, sustainable cities and communities, responsible consumption and production, climate action, life below water, life on land, and peace, justice and strong institutions, and partnerships for the goals (UNDP, n.d.).

The SDGs replaced the Millennium Development Goals (MDGs). The Millennium Development Goals (MDGs) had previously started a global effort in 2000 to tackle the indignity of poverty. The MDGs established universally-agreed and measurable objectives for preventing deadly diseases, tackling extreme poverty and hunger, and expanding primary education to all children, among other development priorities.

For 15 years, the MDGs drove progress in a number of important areas: providing much needed access to water and sanitation, driving down child mortality, reducing income poverty, and drastically improving maternal health. They also launched a global movement for free primary education, moving countries to spend in their future generations. Most notably, the MDGs made huge progress in combating HIV/AIDS in addition to other treatable diseases for instance malaria and tuberculosis (UNDP, n.d.).

The SDGs are a bold commitment to tackle some of the more pressing challenges faced by the world today. All 17 Goals interconnect, which means success in one goal, affects success for others. Dealing with the dangers of climate change effects how we manage our fragile natural resources, better health or achieving gender equality helps eradicate poverty, and nurturing peace and inclusive societies will decrease inequalities and aid in economies prosper.

The SDGs coincided in 2015 with another historic agreement achieved at the COP21 Paris Climate Conference. In addition to the Sendai Framework for Disaster Risk Reduction, signed in Japan in March 2015, these agreements offer a set of achievable targets and common standards to manage the risks of climate change and natural disasters, reduce carbon emissions, and to build back better after a crisis.

The SDGs are unique given that they cover issues that affect everyone. They reiterate the international commitment to permanently end poverty, everywhere. They are resolute in making sure nobody is left behind. More importantly, they engage us all to build a safer, more sustainable, more prosperous planet for all humanity (UNDP, n.d.).

CONCERN FOR GLOBAL ENVIRONMENTAL SUSTAINABILITY

Earlier, environmental issues were not given priority in international relations and were generally considered a domestic issue that required domestic technical solutions rather than multilateral political solutions. The beginning of the twenty-first century witnessed developments towards raising awareness on environmental issues and of their relationship with development, under the notion of "sustainability", and in establishing a multilateral framework including the state and non-state actors to realize this goal of sustainability.

ENVIRONMENTAL SUSTAINABILITY EFFORTS AS DOCUMENTED THROUGH THE YEARS

An Essay on the Principle of Population in 1798 was written by Thomas Malthus. In his seminal work, Malthus stated the potential for geometric increase in population. "The power of population is indefinitely greater than the power of the earth to

Introduction

produce subsistence for man" (Malthus, 1998; Santen, 2013; World Economic Forum [WEF], 2013).

Brief accounts of the major landmarks in the context of environmental protection have been given in the following chronologically since 1948.

The International Union for the Protection of Nature was founded in 1948 with an objective of promoting a unique collaboration of government and nongovernment organizations.

In 1949, the first major meeting of the United Nations on the subject, The United Nations Scientific Conference on the Conservation and Utilization of Resources was held in Lake Success.

"The Challenge of Man's Future" by Harrison Brown was published in 1954 which developed themes that were covered after 25 years by the term "sustainable development."

In 1958, The United Nations Conference on the Law of the Sea was held. It approved draft conventions on environmental protection.

The World Wildlife Fund, later, the World Wide Fund for Nature, was established in 1961 (ADB, 2012).

"Silent Spring" by Rachel Carson was published in 1962. It brought together research on epidemiology, toxicology and ecology to proposition that agricultural pesticides are growing to catastrophic levels, related to damage to human health and animal species. Many consider the release of this book as a turning point in our knowledge of the interconnections among the environment, social well-being and the economy. Many milestones have been reached towards the sustainable development since then (International Institute for Sustainable Development [IISD], 2012).

The International Biological Program begins in 1964 to examine "the biological basis of productivity and human welfare" (ADB, 2012).

In 1967, the Torrey Canyon spilled 118,000 tons of crude oil off Land's End; therefore The Environmental Defense Fund was formed to take up legal solutions to environmental damage. The founders went to the court in order to stop the "Suffolk County Mosquito Control Commission from spray dichlorodiphenyltrichloroethane (DDT) in the marshes of Long Island" (ADB, 2012).

The East European Committee was also set up in 1967 by the Commission on Education and Communication. At the east of the "Iron Curtain;" it was the first and only internationally established nature conservation body.

Paul Ehrlich publishes the "Population Bomb" in 1968. It discusses the connection between resource exploitation, human population, and the environment. In the same year, The Club of Rome was established. It was established to commission a study of global proportions to model and analyze the environmental damage, the dynamic interactions between industrial production, food consumption, population, and natural resource usage (ADB, 2012).

Also in 1968, the Intergovernmental Conference of Experts on the Scientific Basis for Rational Use and Conservation of the Resources of the Biosphere was held in Paris (ADB, 2012).

Intergovernmental Conference for Rational Use and Conservation of the Biosphere (UNESCO) held early discussions on the concept of ecologically sustainable development (IISD, 2012).

Furthermore, the Human Environment Conference was authorized to be held in 1972 by The United Nations General Assembly.

The United Nations Educational, Scientific and Cultural Organization convened the Biosphere Conference in Paris in 1968; it recommended vigorous efforts by public and private organizations and national and international agencies "to establish natural areas for the preservation of species, their habitats, and representative samples of ecosystems" (ADB, 2012).

In 1969, Friends of the Earth was formed as a nonprofit advocacy organization aimed to protecting the planet from preserving biological, cultural, and ethnic diversity; environmental degradation; and empowering citizens to have an influential say in decisions affecting the quality of their lives and their environment (ADB, 2012; IISD, 2012).

Also in 1969, The Pearson Commission on International Development investigated the efficiency of the World Bank's development assistance in the 20 years to 1968. It was the first international commission to use a new approach to development, focused on knowledge and research in developing countries; it leads to the establishment of the International Development Research Centre (ADB, 2012).

The First Earth Day was proclaimed in San Francisco in 1970. It was held as "a national teach-in on the environment." Approximately 20 million people participated in peaceful demonstrations across the United States (ADB, 2012; IISD, 2012).

The International Development Research Centre was established in that year that aimed to support research that promotes development and growth in developing countries (ADB, 2012).

In 1971, Greenpeace begins operations with a program to stop environmental damage by way of civil protests and nonviolent interference. Moreover, The Founex Report, prepared by the panel of experts, called for "integration of environment and development strategies" (ADB, 2012; IISD, 2012).

The United Nations Educational, Scientific and Cultural Organization established in 1971, the Man and the Biosphere Programme in order to promote sustainable use of natural resources and interdisciplinary approaches to research, education and management in ecosystem conservation. The Council of the Organisation for Economic Co-operation and Development presses in 1971 for adoption of the Polluter Pays Principle, which means that those responsible for pollution should pay the costs. Also in 1971, The International Institute for Environment and Development

Introduction

was established to look for ways for countries to make economic progress without damaging the environmental resource base (ADB, 2012).

René Dubos and Barbara Ward published "Only One Earth" in 1971 which discussed the impact of human activity on the biosphere. Nevertheless, it also expressed optimism that a collective concern for the world could lead humanity to create a common future (ADB, 2012; IISD, 2012). Furthermore, the first-ever International Youth Conference on the Human Environment also gathered in 1971 in Hamilton (ADB, 2012).

The United Nations Conference on the Human Environment was held in Stockholm in 1972. This conference was rooted in the problems of acid rain and pollution of northern Europe. "In our time, man's capability to transform his surroundings, if used wisely, can bring to all peoples the benefit of development and the opportunity to enhance the quality of life. Wrongly or heedlessly applied, the same power can do incalculable harm to human beings and human environment." Also further, "To defend and improve the human environment for present and future generations has become an imperative goal for mankind." It lead to the establishment of United Nations Environment Programme and various national environmental protection agencies (ADB, 2012; IISD, 2012; Santen, 2013; WEF, 2013).

Also in 1972, The Club of Rome published "The Limits to Growth" which predicted the outcomes if population growth is not slowed down; it called for a state of global equilibrium. Northern countries criticized the report for missing the technological solutions; Southern countries were riled because it advocates the neglect of economic development (ADB, 2012; IISD, 2012; Santen, 2013; WEF, 2013; Meadows et al., 2005).

Edward Goldsmith and Robert Allen published "Blueprint for Survival" in the same year. It warned of the irreversible disruption of life-supporting systems on the earth and the breakdown of society and called for a steady-state society. It was signed by more than 30 leading scientists (ADB, 2012).

In 1973, The United Nations Conference on the Law of the Sea was held in New York. It covered navigation, limits, archipelagic status and transit regimes, continental shelf jurisdiction, exclusive economic zones, deep seabed mining, scientific research, protection of the marine environment, the exploitation regime, and settlement of disputes.

The European Environmental Action Programme was also launched in 1973. This was the first attempt to make a sole environmental policy for the European Economic Community (ADB, 2012).

The United States enacts the Endangered Species Act in 1973 which made US one of the countries to apply legal protection for wildlife, fish and plants. The act was enacted to better safeguard, for the benefit of the entire masses, the land's heritage in wildlife, fish, and plants (ADB, 2012; IISD, 2012).

Introduction

The Chipko Movement was born in India in 1973, in response to environmental degradation and deforestation. In this movement, women actions influenced forestry practices and women's participation in environmental issues (ADB, 2012; IISD, 2012).

The United Nations Convention on International Trade in Endangered Species of Wild Fauna and Flora was opened for signature in 1974 in Washington, DC. It was a key step in controlling illegal commerce in furs, ivory, and other products of endangered species. The first World Population Conference also took place in the same year in Bucharest with 135 participant countries. A symposium in Cocoyoc in 1974 identifies misdistribution of resources as a major factor in environmental degradation. The meeting called for development action directed on fulfilling basic human needs. The World Food Conference was held in 1974 in Rome. It laid the foundation for a strategy to attack the world food problem and lead to the creation of the World Food Council and the World Food Programme (ADB, 2012).

Mario Molina and Frank Rowland issued seminal work on chlorofluorocarbons (CFCs) with implications for the ozone layer in scientific journal "Nature" in 1974. Their work calculated that the continued use of CFCs at existing rates would seriously deplete the ozone layer. The Bariloche Foundation published "Limits to Poverty" also in 1974. It called for growth and equity in the developing world. It was the South's reaction to "The Limits to Growth" (ADB, 2012; IISD, 2012).

The Worldwatch Institute was established in 1975 to raise public awareness regarding global environmental threats and catalyze valuable policy responses. In 1984, it started publishing the yearly "State of the World". The Convention on International Trade in Endangered Species of Wild Flora and Fauna was also enforced in 1975 (ADB, 2012; IISD, 2012).

The Dag Hammarskjöld Foundation published "What Now: Another Development" in 1975. It was developed on the Founex and Cocoyoc documents and called for development in poor countries that is endogenous, needs-oriented, self-reliant, ecologically sound, self-reliant, and based on participation and self-management (ADB, 2012).

The United Nations Conference on Human Settlements was held in Vancouver in 1976. It was the first global meeting to link human settlement and environment (ADB, 2012; IISD, 2012).

The United Nations Conference on Desertification was held in 1977 in Nairobi. The United Nations Water Conference was also held in the same year in Mar del Plata. It sets the objective of providing adequate sanitation and clean water to everyone in the world by 1990 (ADB, 2012).

Introduction

The Organisation for Economic Co-operation and Development had in 1978, resumed research on economic and environment linkages (ADB, 2012).

In 1979, The Convention on Long-range Transboundary Air Pollution was adopted. Moreover, Robert Stein published "Banking on the Biosphere" which reviewed the practices and procedures of nine multilateral development agencies. The World Climate Conference was also held in 1979 which concluded that the "greenhouse effect" from increased rise of carbon dioxide in the atmosphere demands necessary international action (ADB, 2012).

In 1980, The International Union for Conservation of Nature published the "World Conservation Strategy" in which, the section "Towards Sustainable Development" identified the main factors of habitat destruction as population pressure, poverty, trading regimes and social inequity; the strategy demanded a new international development strategy to balance out inequities. The term "sustainable development" was first introduced by the World Conservation Strategy in 1980, into the international policy debate. The Independent Commission on International Development Issues published "North–South: A Program for Survival" in 1980. It is also known as the Brandt Report which called for a new economic relationship between North and South. The United States Department of State and the Council on Environmental Quality released "The Global 2000 Report to the President;" it recognized biodiversity for the first time as a crucial factor to the accurate functioning of the planetary ecosystem (ADB, 2012; IISD, 2012).

In 1981, The World Health Assembly of the World Health Organization adopted a Global Strategy for Health for All by the Year 2000. It asserts that the major social goal of governments should be to attain a level of health by all individuals that would permit them to lead economically and socially productive lives (ADB, 2012).

The UN Convention on the Law of the Sea set a comprehensive framework for ocean use in 1982 and outlined provisions on pollution prevention, ocean conservation, and restoring and protecting species populations (Worldwatch, n.d.; ADB, 2012).

In 1982, the United Nations World Charter for Nature adopted the principle that all form of life is special and should be respected despite of its value to humankind; it called for an acceptance of our dependence on natural resources and the requirement to control their exploitation (ADB, 2012).

The U.S. National Academy of Sciences and the U.S. Environmental Protection Agency issued reports in 1983 concluding that the increase in carbon dioxide as well as other greenhouse gases in the Earth's atmosphere will likely result in global warming (Worldwatch, n.d.).

Introduction

The World Commission on Environment and Development was formed in 1983. It was chaired by Gro Harlem Brundtland. The commission worked for 3 years to compile together a report on economic, social, cultural, and environmental issues.

Development Alternatives was also established in 1983 in order to foster a new relationship between environment, technology, and the people to achieve the goal of sustainable development (ADB, 2012).

The Third World Network was established in 1984 to serve as the advocate voice of the South on issues of environment, economics, and the development (ADB, 2012).

The International Conference on Environment and Economics in 1984 concluded that the economics and environment should be mutually reinforcing (ADB, 2012).

Scientists reported in 1985, the discovery of a "hole" in the Earth's ozone layer, since data from a British Antarctic Survey showed that January ozone levels dropped 10 percent below the ozone levels of the previous year (ADB, 2012; Worldwatch, n.d.).

In 1987, UN World Commission on Environment and Development highlighted the challenge of bringing several billion into the mainstream economy along with the imbalances in consumption.

Given population growth rates, a five- to tenfold increase in manufacturing output will be needed just to raise developing world consumption of manufactured goods to industrialized world levels by the time population growth rates level off next century. Later in the report it is pointed out with prescience that, 'Perceived needs are socially and culturally determined, and sustainable development requires the promotion of values that encourage consumption standards that are within the bounds of the ecologically possible and to which all can reasonably aspire. (WEF, 2013; World Commission in Economic Development, 1987)

In 1992, UN Conference on Environment and Development concluded that the unsustainable pattern of consumption and production is the main cause of the constant deterioration of the global environment, specifically in industrialized countries, which is an issue of grave concern, exacerbating poverty and imbalances (United Nations, 1992; WEF, 2013).

In 2002, the closing statement of World Summit on Sustainable Development called for countries to "Encourage and promote the development of a framework ... in support of regional and national initiatives to accelerate the shift towards sustainable consumption and production to promote social and economic development within the carrying capacity of ecosystems" (WEF, 2013; United Nations, Report of the World Summit on Sustainable Development, 2002).

Introduction

In 2003, Marrakech Process on Sustainable Consumption and Production was launched. This UN process was commenced at the "first international expert meeting on the 10-year framework held in Marrakech, Morocco, organized by the UN Department of Economic and Social Affairs (UNDESA) Division for Sustainable Development and the UN Environment Programme (UNEP)" (WEF, 2013). For almost a decade, an alliance of willing countries has been working to promote sustainable production and consumption, particularly through policy guidelines and in emerging economies (WEF, 2013; United Nations Environment Programme, Division of Technology, Industry and Economics, 2011).

UN Conference on Sustainable Development (Rio+20) was held in 2012. "We adopt the 10-Year Framework of Programmes (10YFP) on sustainable consumption and production ... We invite the UN General Assembly ... to take any necessary steps to fully operationalize the framework." After about a decade of moving ahead without formal agreement by all UN countries, the Marrakech Process 10-year framework was finally adopted being one of the few accomplishments of a controversial Rio+20 Summit (WEF, 2013; United Nations, 2012).

ENVIRONMENTAL SUSTAINABILITY AND INTERNATIONAL COOPERATION

International Relations are very important aspect in considering the options or directions of development in a country. Therefore, strengthening or increasing a state's relations with other states in the world are its foremost priority and the same has been kept in view while formulating its foreign policy. The era of Globalization started with a promise to integrate world nations ending discrimination and deprivation of the developing countries and developing the entire globe by strengthening international cooperation and hence, international relations. Thus, the world became a global village where the countries are now determined to cooperate with each other realize global agendas and help implement those policies to achieve such common goals. These agendas or goals include Environmental Sustainability.

The issue of global warming is very alarming for the world and all the states feel their responsibility to contribute to address this issue; however, greater cooperation is needed in terms of finding a helping hand.

Earlier, environmental issues were not given priority in international relations and were generally considered a domestic issue that required domestic technical solutions rather than multilateral political solutions. The beginning of the twenty-first century witnessed developments towards raising awareness on environmental issues and of their relationship with development, under the notion of "sustainability", and in establishing a multilateral framework including the state and non-state actors to realize this goal of sustainability.

International cooperation is useful to force the reluctant elements and implement the necessary steps to stop environmental degradation which maybe happening partly because of development goals for instance, factory and industry establishment, high and unchecked energy consumption through vehicles, industries as well as through domestic and commercial use. Therefore, international pressure has been helpful in promoting environmental approach to development for example, bio-fuel, choice of raw material for electricity production and for other manufacturing products along with emphasizing on importance of recycling.

There have been many world summits in an effort to reduce environmental degradation and make a collective effort in reaching a globally ratified agreement to work towards environmental sustainability.

However, international involvement is also hindering the progress in environmental sustainability to realize their economic policies resultantly such goals couldn't be achieved. This has changed the international relations in essence and spirit which has shifted from once being of cooperation and coordination nature to competition among the states. The book includes some case studies of different countries in addition to research on patterns of consumption, international business and world economies.

This book looks to discuss and address the difficulties and challenges of achieving environmental sustainability. The chapters in this book address different aspects of international relations, ranging from foreign policies, cooperation and competition, conflict and conflict resolution, trade, business, technology, economics and politics and related topics concerning environmental sustainability. Additionally, the book will explore the changing world situation in the contemporary globalized world with some states exerting their influence subsiding the spirit of cooperation and overlooking the issue of environmental sustainability.

Practitioners and scholars from diverse regions of the world have contributed to this book. These experts in their field have provided account of the issues, solutions and opportunities in terms of the environmental sustainability and cooperation at a global level.

Introduction

OBJECTIVE OF THE BOOK

This comprehensive and timely publication aims to be an essential reference source, building on the available literature in the field of international relations in the contemporary globalized world while providing for further research opportunities in this dynamic field. It is hoped that this text will provide the resources necessary for policy makers, academicians, researchers, government officials and international institutions to adopt and implement measures to make better the existing situation.

TARGET AUDIENCE

Policy makers, academicians, researchers, advanced-level students, government officials and international institutions will find this text useful in furthering their research exposure to pertinent topics in international relations and assisting in furthering their own research efforts in this field.

Sofia Idris
Independent Researcher, Pakistan

REFERENCES

Asian Development Bank (ADB). (2012). *World Sustainable Development Timeline*. Retrieved from https://www.researchgate.net/profile/Olivier_Serrat/publication/266878643_World_Sustainable_Development_Timeline/links/543e3e430cf2d6934ebd20af/World-Sustainable-Development-Timeline.pdf?origin=publication_list

International Institute for Sustainable Development (IISD). (2012). *Sustainable Development Timeline*. Retrieved from https://www.iisd.org/pdf/2012/sd_timeline_2012.pdf

Malthus, T. (1998). *An Essay on the Principle of Population*. London: Electronic Scholarly Publishing Project. Available at: http://129.237.201.53/books/malthus/population/malthus.pdf

Meadows, D. (2005). *Limits to Growth: The 30 year Update*. London: Earthscan.

Morelli, J. (2011). Environmental Sustainability: A Definition for Environmental Professionals. *Journal of Environmental Sustainability*, *1*(1), 1–10. doi:10.14448/jes.01.0002

United Nations. (1992). *United Nations Sustainable Development, Earth Summit: Agenda 21, Chapter 4.3*. Available at: http://www.un.org/esa/sustdev/documents/agenda21/english/Agenda21.pdf

United Nations. (2012). *The Future We Want*. Available at: http://www.uncsd2012.org/content/documents/727The%20Future%20We%20Want%2019%20June%201230pm.pdf

United Nations Development Programme (UNDP). (n.d.a). *Sustainable Development Goals: Background on the Goals*. Retrieved from http://www.undp.org/content/undp/en/home/sustainable-development-goals/background.html

United Nations Development Programme (UNDP). (n.d.b). *Sustainable Development Goals*. Retrieved from http://www.undp.org/content/undp/en/home/sustainable-development-goals.html

United Nations Environment Programme, Division of Technology, Industry and Economics. (2011). *Paving the Way for Sustainable Consumption and Production: The Marrakech Process Progress report*. Available at: http://www.unep.fr/scp/marrakech/pdf/Marrakech%20Process%20Progress%20Report%20FINAL.pdf

United Nations, Report of the World Summit on Sustainable Development. (2002). Available at: http://www.johannesburgsummit.org/html/documents/summit_docs/131302_wssd_report_reissued.pdf

van Santen, L. (2013). *Timeline: a brief history of sustainable consumption*. World Economic Forum. Retrieved from: https://www.weforum.org/agenda/2013/11/timeline-a-brief-history-of-sustainable-consumption/

World Commission in Economic Development. (1987). *Report of the World Commission on Environment and Development: Our Common Future*. Available at: http://www.un-documents.net/our-common-future.pdf

World Economic Forum (WEF). (2013). *Sustainable Consumption: Stakeholder Perspectives*. Retrieved from: http://www3.weforum.org/docs/WEF_ENV_SustainableConsumption_Book_2013.pdf

Worldwatch. (n.d.). *Environmental Milestones: A Worldwatch Institute timeline tracing key moments in the sustainability movement from 1960s to 2004*. Retrieved from: https://www.worldwatch.org/brain/features/timeline/timeline.htm

Section 1
Globalization and Global Governnce

Chapter 1
Globalization and Rethinking of Environmental Consumption From a Sustainability Perspective

Luke A. Amadi
University of Port Harcourt, Nigeria

Prince Ikechukwu Igwe
University of Port Harcourt, Nigeria

ABSTRACT

Sustainable environmental consumption has been a marginalized concept in international development studies and cooperation. In recent decades, there has been growing interest in identifying robust indicators that demonstrate the evidence of globalization and unsustainable environmental consumption. Globalization is premised on integrating the world into a global village. Various dimensions of globalization have different effects on the ecosystem. Plausible evidence linking globalization trajectories into practical interactions suggesting sustainable environmental consumption has been less lucid as the effects of globalization on the ecological environment does not provide clear patterns. This hugely significant problem has reopened critical debates on novel thinking on dynamics of environmental consumption patterns of the affluent societies in the era of globalization and its implications on environmental sustainability. This chapter deployed content analysis methodology and political ecology framework to review and analyze seminal studies on sustainable consumption and globalization, including relevant globalization indexes. The aim is to provide evidence of the impact of globalization on environmental consumption. The chapter suggests that globalization results in asymmetrical and

DOI: 10.4018/978-1-5225-3990-2.ch001

deleterious natural resource extraction between the affluent North and poor South. It offered alternative thinking in which sustained policy framings and international development collaboration could be institutionalized to strengthen sustainable environmental consumption one which is premised on ecological justice and natural resource equality.

INTRODUCTION: GLOBALIZATION AND ENVIRONMENTAL SUSTAINABILITY INERTIA

Environmental consumption appears to be one of the most pressing sustainability challenges of our times particularly in the era of globalization. The evidence suggesting how globalization undermines sustainable environmental consumption has been scant and less clear. Despite the debates on ecological footprints which provide evidence that humanity's environmental footprint is highly un-sustainable and that radical changes in the global human organization are necessary (Hoekstra & Wiedmann, 2014). It appears the patterns of consumption in the ongoing globalization point to the opposite direction, this stimulates research curiosity.

Globalization and its effects have different interpretations by different stakeholders. For instance, in the global climate change dialogue, *Eco Watch* a leading US environmental news site reported in the month of March,2017 that Scott Pruitt, the new head of the U.S. Environmental Protection Agency (EPA), does not think that carbon dioxide is a "primary contributor" to climate change—even though the actual science says it is (Chow, 2017).However, most scientists – especially those working in the Intergovernmental Panel on Climate Change (IPCC) – believe that increases in emissions of carbon dioxide (CO_2) from human activity are the primary cause of global warming(OECD, 2013).

There are several perspectives and definitions of globalization, Rennen and Martens (2003) posit that globalization is the intensification of cross-national interactions that promote the establishment of trans-national structures and the global integration of cultural, economic, ecological, political, technological and social processes on global, supra-national, national, regional and local levels. Studies linking environmental consumption to globalization provide a number of useful insights. Ecologist, Katrina Rogers (1995) linked ecological breakdown to multinational corporations which have ecological security implications. Rogers(1995)argued that "ecological security refers to the creation of a condition where the physical surroundings of a community provide for the needs of its inhabitants without diminishing its natural stock".

Similarly, scholars of environmental consumption have come close circle with the effects of consumption on the environment. Stern (1997) explored the environmental impacts of consumption and argued that it is "rooted partly in environmental high politics". He examined a broad disciplinary definition of consumption and noted the particular case of ecological perspective which builds on Vitousek, et al., (1986) who argued that any organism that obtains its energy by eating is a consumer. Human consumption corresponds to what humanity does with the estimated 40 percent of global terrestrial net primary productivity (NPP) that we "appropriate". Stern (1997) provides a broader definition which emphasizes the "environmental impact of human choices and actions" (rather than of consumption alone) as the object of research and demonstrates that choices and actions which affect the environment are integral to environmental consumption.

Beyond this, how globalization gives rise to unsustainable consumption remains less clear and perhaps a new research agenda. This forms part of the central focus of this chapter. The chapter follows Stern (1997) who argued that consumption consists of human and human-induced transformations of materials and energy and that consumption is environmentally important to the extent that it makes materials or energy less available for future use, moves a biophysical system toward a different state or through its effects on those systems, threatens human health, welfare or other things people value. It also builds on Figge, Oebels & Offermans (2017) who provided evidence of the effects of globalization on ecological footprints. The aim is to substantially provide deepened evidence of the impact of globalization on environmental consumption. The evidence will illustrate the analytical utility of our theoretical framework which is the Marxian political ecology approach. The chapter demonstrates that research tools generated through this line of inquiry could be useful for a multi-level explanation of the basis for rethinking unsustainable consumption in the era of globalization.

Sustainable consumption means the provision of services and related products, which respond to basic needs and bring a better quality of life, while minimizing the use of natural resources and toxic materials as well as the emissions of waste and pollutants over the life cycle of the service or products, with a view not to jeopardize the needs of future generation (UNDP,1998).

The chapter argues that global modes of environmental consumption and accumulation are intimately linked to environmental degradation (e.g. extraction of natural resources and multiple forms of pollution via commodity production) (Jorgenson & Kick, 2003).The central objective of the chapter is to demonstrate how globalization has resulted in unsustainable natural resource consumption which negatively impacts the global South and how policy choices to transform the ecological challenges could be institutionalized.

These will be explored through the linkages between globalization and unsustainable consumption. This includes the rise in global consumer culture, patterns of consumption of the high-income societies, the trend toward increased ecological breakdown and how this fits into the contemporary globalization debate.

The rest of the chapter is structured as follows; background, theoretical framework, research methodology, conceptual issues, globalization, unsustainable environmental consumption and Marxian political ecology debate, recommendations, future research directions and conclusion.

BACKGROUND

Global attention on environmental sustainability was rekindled in the Rio summit of 1992 at the first largest world summit on environment held in Rio De Jenerio. Its Agenda 21-*the Plan of Action for Implementation of Sustainable Development* sparked novel curiosity on environmental sustainability. Decades following the summit have seen the emergence of environmental sustainability as a distinct field of inquiry among scholars and policy makers both at the local and international levels. Its linkages with globalization perhaps jostle for scholarly attention.

The epochal globalization started with a promise to integrate world nations, ending discrimination and deprivation of the developing countries and developing the entire globe by strengthening international cooperation and hence, international relations. Thus, the world became a global village where nations are linked with each other. Increasing scholarly curiosity had emerged in recent decades on possible nexus between globalization and environmental sustainability (Miller, 1995; Rogers,1995; Figge, Oebels & Offermans, 2017), as several development decades point to perverse environmental disaster.

Miller (1995) argued that 'developing countries have lost leverage in global environmental politics, as result of the end of the Cold War'. The varying development experiences and disparities between the North and South, the resurgent global energy crisis arising from unsustainable energy consumption, the imminent dangers of global warming and uncertainties of climate change, the emergence of global water politics, in particular, the rise in global water corporations and water commodification, results in the danger of global fresh water shortages. Worldwide, about 2.3 billion people suffer from diseases that are linked to water problems (Kristof, 1997; UN, 1997; WHO, 1997).

Beyond water, there are related ecological problems linked to globalization. This includes food insecurity arising from unequal access and unsustainable food consumption associated with the rise in global food chains, stores, networks and re-occurring issues of genetically engineered (GE) foods. This involves cutting parts of deoxyribonucleic acid (DNA) out of one organism and inserting them into another using a virus to enter the cell membrane, and an antibiotic gene maker (Haley, Hatt & Turnstall,2005). Jeffrey Smith of the Institute of Responsible Technology had provided evidence which suggests the connections between Genetically Modified Organisms (GMOs) and cancer. Smith (2017) among a number of points argued that several cancer rates in the US are rising in parallel with increased use of glyphosate on GMO soy and corn fields. These include leukemia and cancers of the liver, kidney, bladder, thyroid, and breast. Food expert, Timothy Wise (2015) identified "feeding illusions" linked to agribusiness, the bleak future of food and the elusive promises of globalization. This is further elaborated in the Berne Declaration (2013) report which states that corporate lobbyists work in government institutions and protect their corporate interests on food standards, approval of pesticides, GM seeds, trade agreements, or the public research agenda. This reinforces the issues of unsustainable food consumption as Terhune (2004) argues that the corporations that produce this food admit that they have no moral obligation to monitor diet, and indeed, they do all they can to promote consumption and profits. Schor (2005) points to the disproportionate consumption of the affluent societies particularly the United States and its ecological implications. American economist, Lesther Thurow (1992) argued that the effects of unsustainable environmental consumption is no longer in doubt. Shiva (1988) argues that natural resources have become a commodity and this inevitably results in resource wars and global crisis. This is evident among the poor societies of the Third World who subsist on less than $2dollars per day (World Bank, 2015). For instance, in Latin America, Roberts and Thanos (2003) argue that to understand the environmental crisis in Latin America requires an understanding of poverty and inequality in the entire region. This has been in contrast with the affluence and unsustainable consumption patterns of the high income societies (Hobson,2003; Schor, 2005). Similar concern on globalization and the environment is raised in Asia (Esty & Pangestu, 1999).

Against this background and following the increasing awareness on unsustainable consumption and divergent turn globalization had taken including environmental politics and uncertainties of climate change, there is need for natural resource equity and ecological justice. This chapter is concerned with unsustainable consumption of environmental resources in the era of globalization. It puts forward sustainability analysis of consumption one which demonstrates the need for preservation of human and environmental components of nature.

Figure 1 shows global annual average temperature measured over land and oceans. Red bars indicate temperatures above and blue bars indicate temperatures below the 1901-2000 average temperature. The black line shows atmospheric carbon dioxide concentration in parts per million.

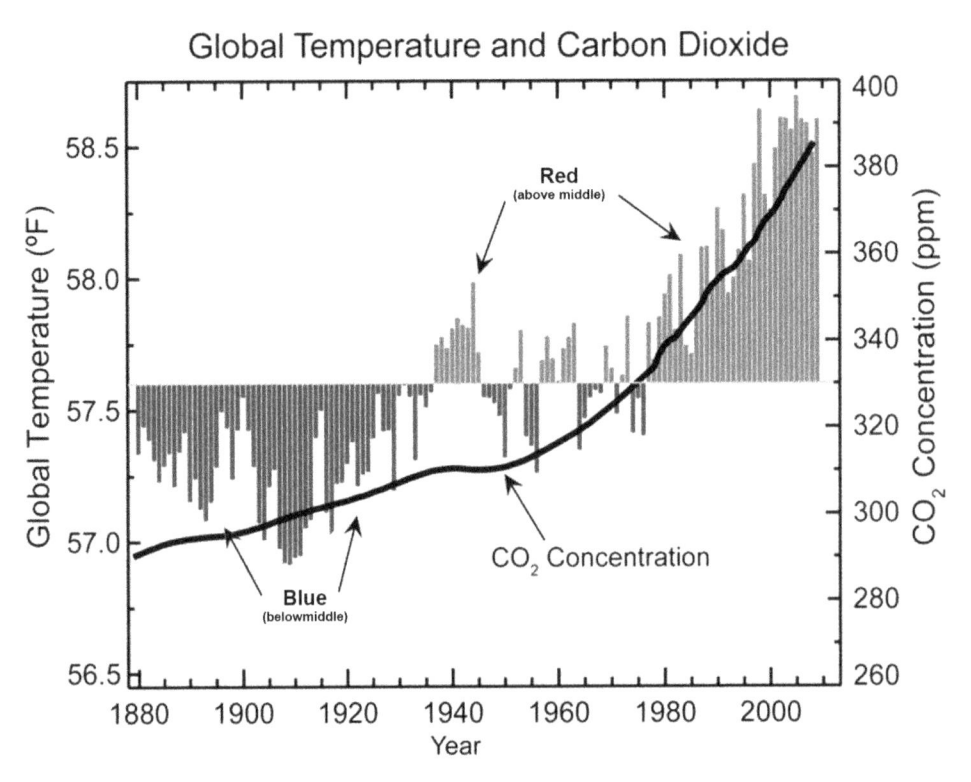

Figure 1. Global annual average temperature measured over land and oceans
Source: *United States National Climatic Data Center – NOAA*

THEORETICAL FRAMEWORK

Theoretical analysis on globalization often proceeds in a way which fails to substantially explore the facets of global inequality. This does not adequately account for the realities of globalization such as the implications of unsustainable consumption in global contexts. For instance, consumption disparities among the underdeveloped and industrialized societies largely remain unresolved in research and policy framing. On the contrary, globalization allegedly emphasizes the interconnected and overlapping nature of societies as globalization presupposes the disappearance of boundaries (Ohmae, 1995; Rosenau, 1996; Gidens, 1999). Such superficial conceptions could be misleading.

Several studies on globalization and consumption rely on relatively problematic frameworks which do not lucidly capture the effects of ecological injustice in the patterns of consumption between the rich and poor societies. Nor critically investigate the complexities of globalization in the domain of sustainable and equitable development. Ferguson (2006) recounts that globalization divides the entire social space into discrete, unequal domains. Such spheres of division results in complex theoretical and conceptual difficulties. For example, within the technological domain, Castells (2000) argues that the internet hypertext had emerged as a central motive force in the "new society" or what he termed the "network state". Within the cultural thesis, Amadi and Agena (2015) and proponents of cultural imperialism contend that there is subtle but persistent culture dislodgement arising from unequal culture contacts between the dominant and the recessive culture.

The economic component centers on "global economic integration" in which the affluent societies integrate and appropriate the resources of the less developed societies through market relationships (Stiglitz, 2003, 2016). Similarly, there are studies concerned with the need to redress the relationship between economic institutions, such as corporate production, on the one hand, and non-economic institutions, such as changing patterns of consumption, institutional roles or political structures. The complexity of such roles is linked to imperialism imbedded in globalization such as the rise in Western multinationals and the Bretton Woods institutions such as the International Monetary Fund (IMF) and the World Bank. This underpins similar accounts that explores the influence of Western corporation (Korten, 1995) and the rise in new imperialism (Calinicos, 2009). The ecological component of resource insecurity has equally been foreshadowed (Rogers, 1995). In a related account, Harvey (2005) points to the complex economic and geopolitical nature of imperialism arguing that the neoliberal development machination is antithetical to the environment.

While globalization plays multidimensional roles in ecological studies, sustainable environmental consumption and deleterious resource use have been less conceptualized. Similarly, the ecosystem which is composed of life support system including human and non- human species (Goodland, 1995) is rarely explored in relation to globalization trajectories.

Thus, environmental consumption and globalization nexus is suitably examined from the lens of the Marxian Political ecology. The political ecology framework explores patterns and dynamics of "resource inequality and exploitation" particularly in the Third World (Bryant, 1998). The dialectical exploration of this inherent inequality makes the Marxian political ecology the most suitable framework to examine globalization and unsustainable environmental consumption. Amadi and Igwe (2016) argue that political ecologists seek to explore the critical challenges posed by poor and unjust natural resource management. Ecological Marxists argue for a more equitable and efficient mode of natural resource extraction and indeed, the management of the "free" markets and private property rights considered part of "the economy" (Blaikie & Brookfield, 1987; Amadi & Igwe, 2016).

Understanding the Marxian political ecology helps in building a framework to comprehend and analyze inequality inherent in capitalist exploitation, the implications to humanity and the environment. Ecological Marxism which is a critique of the neo-Malthusian theory examines the primacy of inequality and deleterious implications in the interactions between humans and their natural environment. Political ecology underscores a novel engagement of stakeholders to create a condition where the environment could be consumed equitably without diminishing its natural stock. Political ecology assumes that political policies should influence ecological choices and that human survival largely depends on modalities in which the environment is used and protected.

There are a number of reasons for the suitability of this framework first is its emphasis on inequality in natural resource use and again its relevance in understanding prevailing asymmetrical consumption patterns of the high-income societies in relation to globalization. This framework offers a new way of conceptualizing the linkages between globalization and environmental consumption (see Table 1).

RESEARCH METHODOLOGY

Studies on globalization and environmental consumption are riddled with ambiguity and complexities. This study is a content review and analysis of relevant seminal data. Content review is a suitable methodological approach to the study of globalization and its effects on consumption as it provides a deepened understanding of the causal evidence which suggests the impact of globalization on unsustainable environmental consumption.

As a step in identifying the evidence, we first undertook an assessment of the concept of globalization, robust indicators of effects of globalization on unsustainable environmental consumption, a review and analysis of relevant authentic literature including mapping and analysis of data already available in the field of inquiry.

The assessment of globalization and environmental consumption was followed by conceptual issues and identification of the salient ecological problems linked to unsustainable resource consumption. A balanced analysis was sought through an extensive review of pro globalization scholarship linked to the ecological modernization research (Spaargaren & Mol, 1992; Mol, 2001) and anti-globalization debate aimed at environmental sustainability (Korten, 1995; Stiglitz, 2003; Davisdon & Hatt, 2005; Hrvey, 2005). The aim is to understand the concerns raised on globalization and ecological issues from both pejorative and commemorative perspectives.

A review of the literature on environmental resource consumption and global North /South divide, including resurgent ecological problems was undertaken. The study deployed the Marxian political ecology theory which examines the logic of inequality in natural resource extraction.

Table 1. Key Global Environmental Trends/Summits 1970-2012

Year	Summit/ Environmental Development
1970	The Club of Rome and limits to growth document
1971	The Convention on Wetlands of International Importance especially as Waterfowl Habitat is adopted in Ramsar, Iran
1972	The UN Conference on the Human Environment is held in Stockholm, Sweden
1972	The United Nations Environment Programme (UNEP) is established, with its headquarters in Nairobi, Kenya
1972	The Convention Concerning the Protection of the World Cultural and Natural Heritage is adopted in Paris, France
1972	The Convention on the Prevention of Marine Pollution by Dumping of Wastes and other Substances is adopted in London, United Kingdom, and Mexico City, Mexico
1973	The 'oil weapon' is first used on the world oil market by the Arab oil exporting countries. This has a devastating impact, especially on the economies of developing countries, including those of Africa
1973	The Convention on the International Trade in Endangered Species of Wild Fauna and Flora (CITES) is adopted in Washington, D.C., USA
1973	The International Convention for the Prevention of Pollution from Ships is adopted in London, United Kingdom
1975	The Convention on Wetlands of International Importance Especially as Waterfowl Habitat enters into force

The specific methodological tool followed the Globalization Indices (GI) which has been a useful tool to monitor, measure and communicate the complexity and multidimensionality inherent in globalization. GIs allow for a relative ranking or comparison of country performance (OECD, 2008) with respect to globalization, on the basis of indicators.

This includes validated indices that attempt to show evidence of ecological impacts of globalization such as the Maastricht Globalization Index (MGI) (Figge & Martens 2014; Martens & Raza 2009). There are also the globalization indexes: *G-Index* by World Markets Research Centre, Globalization Index by A.T Kearney, *Global Index* by Trans Europe research program of the European Science Foundation, Globalization Index by the Centre for Study of Globalization and Regionalization (CSGR)at Warwick University (Lockwood & Redoano, 2005).

Equally, the KOF index (Konjunkturforschungsstelle) which means business cycle research institute (Dreher, 2006; Dreher et al. 2008) was followed. Beyond these, the chapter builds on Figge, Oebels and Offermans (2017) who attempted an empirical exploration of the effects of globalization on ecological footprints and Dreher et al. (2008), Martens, et al. (2010) and Martens and Raza (2010) whose works have at various instances examined both the sustainability and measurement of globalization.

A broader analysis was undertaken including a number of regional studies on the impact of globalization on the Third World in global environmental politics(Milner,1995), environmental crisis in Latin America, Asia and Africa (Roberts & Thanos, 2003; Esty & Pangestu, 1999; Scheren, Ibe, Janssen & Lemmens,2002; Fayiga, Ipinmoroti, & Chirenje, 2017). Assessment based on the relevance of measuring environmental sustainability in ecological contexts (Amadi & Imoh-Itah,2017) were examined.

In the second step, a more detailed analysis of globalization and unsustainable environmental consumption with evidence of ecological disruptions were explored within the Marxian political ecology debate (Korten, 1995; Hart, 1997; Bryant, 1998; Shove, 2003; Hobson, 2003; Davidson & Hart, 2005; Bakan, 2004; Amadi & Igwe,2016; Amadi, Igwe & Ogbanga, 2016). This was aimed at obtaining information and evidence of unsustainable consumption and related deleterious effects on humans and the ecosystem.

Finally, a deepened review of expert studies and seminal institutional literature on globalization and unsustainable environmental consumption was undertaken in line with the over -all objective of the study (UN, 1987; UN, 1997; WHO, 1997; UNDP, 1998; UN, 2002a; UN, 2012; OECD, 2013).

CONCEPTUAL ISSUES

Conceptual clarity on the concepts of "environmental consumption" and "globalization" is important to distinguish between both concepts and to identify possible linkages. The orientation towards globalization and sustainable consumption has given rise to a number of conceptual debates following the ongoing social change necessitated by the logic of globalization.

Essentially, concepts like "environmental consumption" and "globalization" form part of a wider debate deployed in order to understand the processes, practices and social relations which shape human activities and the use of the natural environment. They are also contentious categories as there are challenges in measuring sustainability (Amadi & Imoh-Itah, 2017) and in particular, how globalization is impacting the ecosystem. Also global collaboration and mitigation of deleterious environmental challenges such as gaseous emission, pollution, deforestation, ecological breakdown, climate change vulnerability etc have been less lucid.

Consumption has been deployed to examine a wide range of issues. Bauman (1998) equates consumption to destruction and contends that a thing is destroyed once it is consumed. Regarding globalization and consumption, a number of useful conceptual debates have been advanced. Within cultural imperialism thesis, "mutation" has been conceptualized to identify globalization as instrument of "culture dislodgement" among the developing societies (Amadi & Agena, 2015). Culture mutation provides conceptual thread in understanding the dislodgement of the periphery culture by the Western culture. The argument is that 'cultural imperialism' is reinforced through globalization, resulting in the emergence of "new identity" and new ways of life at variance with traditional African culture and values (Amadi & Agena, 2015). In this context, the dominant culture propagated by the West, dislodges the non-dominant culture of the developing societies.

"Consumption culture" also entails attachment or totality of ways of consumption of a people. It also underscores relational interaction including lifestyle, identity, hegemony, dress code, eating or use of objects or things. Part of the conceptual debates on persistent challenge of unsustainability is attributable to the pursuit of economic benefits by the capitalist societies as they do not prioritize environmental implications. Harvey (2005) argued that inclusive and participatory development have been subverted in the neo liberal order leading to "creative destruction". Thus, there is urgent need for intervention and necessary steps to check environmental degradation which may hamper environmental sustainability. Worzel (1994) in the particular case of persistent pollution argued that what the affluent societies of Europe and North America do not understand is that its either they continue "doing this" or kill all of us one day.

Related conceptual evidence of resource wars and conflicts in the developing societies particularly fostered by the logic of "eco imperialism" abound. A concept popularized by ecologist, Paul Driessen (2003) to address the imminent dangers of capitalist resource extraction by the Western imperialists. Driessen (2003) argued that industrial pollution arising from global oil giants has resulted in diseases in the poor societies. Including the logging of woods and deforestation, running of factories and unchecked environmental risks.

Neoliberalism has given rise to freedom which is often at variance with equitable development (Harvey,2005). The has been an offshoot of imperialism riddled with exploitation (Calinicox, 2009). Proponents of the risk society argue that such scenarios are prone to vulnerability (Beck,1992).

In these conceptual perspectives, environmental sustainability debate aims to represent an advance on political ecology in its conceptualization of the environment as integral component of development, in relation to inclusive, participatory and equitable development practices. Since at least 1992, a number of seminal studies have emerged in the field of environmental sustainability, pointing out its relevance in the over- all development studies, reviewing key trends in its consolidation and research trajectories that could advance the understanding of the ecological implications of globalization and related research gaps. The gaps include prioritizing ecological concerns in natural resource consumption. David Haglund points out that ecological consideration was recognized in international relations more than a quarter century. The implications these changes have on the natural environment is less theorized in policy discourse.

Conceptual debates on globalization have been complex from both pejorative and commemorative strands. The term globalization is not new in both sustainability and ecology studies. Globalization has furthered dynamics of development trends in the international world. McGrew (1992) reinforces the broader scholarly reach of the concept of globalization within the international relations discourse pointing out that globalization intensifies global interconnectedness with a wide range of linkages including the movement of 'goods, capital, people, knowledge, images, crime, pollutants, drugs, fashions and beliefs across territorial boundaries'. The character of the international capitalist system is another important issue. Veseth (2004) recounts the increasing North /South dichotomy which remains a factor in global unequal development. Similarly, Collier and Gunning (1999) posit that the international capitalist system is a system of lawlessness. This suggests the need for order in the patterns of interaction and relationship in the international capitalist system.

Debates linking globalization with sustainable environmental consumption is loosely defined and can represent a myriad interpretation including food and resource networks, communication, technology, environment or the integration of markets. Amidst unsustainable consumption, "Biosecurity" is argued not to have been guaranteed in several countries of the North (Berne Declaration, 2013). For instance, in Germany, one third of antibiotics sold are used in animal production, in China it is one half. In the US, where antibiotics are allowed to accelerate growth, eight times more antibiotics are used in factory farms than in hospitals (Berne Declaration, 2013). The consequences include persistence of antibiotic resistant bacteria and increasing rate of people with infections resistant to antibiotics. The World Health

Organization reports that the scenario remains a serious threat for human health. For instance, when Fluoroquinolones, which is among the commonly used antibiotics was banned from poultry in the US, the manufacturers Bayer increased global sales of "Baytril" by 11% in 2010 (Berne Declaration, 2013).

Globalization and food consumption disparity have been linked to profit maximization by the few affluent societies and exploitation of majority of the low-income societies of the global South. The Berne Declaration (2013) argued that much of what is consumed in the North is produced more cheaply in the global South that the profits are made by only a few, predominantly Northern companies. The big losers they argued are the plantation workers and small farmers in the South, who are the weakest links in the "value chain".

This increasingly informs the resurgent debate on rethinking globalization in sustainability contexts on the grounds that such research agenda has not been adequately explored. This often creates various theoretical and analytical difficulties, particularly in the affluent societies where the sustainability depictions of consumption are less theorized.

Although, the treatment of globalization as a development paraphernalia has never gone unchallenged in environmental sustainability debates. The geographer, Simeon Dalby (2002) argued that "many things with a vague environmental designation now apparently endanger modern modes of life in the North". This informs and underpins the understanding of the challenges of globalization and consumption patterns of the North which has rarely been considered a derivative "effect of unsustainability". Bello (2003) argued that there is need for the poor societies to de-globalize and later re-globalize. This among others reinforces the increasing complexities and contradictions of globalization (Rosenau, 1996; Giddens, 1999). The recently emergent field of environmental sustainability attempts to rectify the deficiency in unsustainable consumption in various ways, and while it succeeds in revaluing unsustainable consumption as it affects environmental sustainability, it fails to resolve the difficulties associated with the underlying challenges of globalization.

Castells (2000) identified the increasing changes in the world in present decade, including changes in the ways business is done and suggests the multi-dimensional effects of globalization. Castells (2000) explores the rise of the network state and points out that the dynamics of international business, technological advancement, response and commitment to global relationship is equally witnessing changes from divergent perspectives. He linked this to technology hypertext and argues that the result or consequences of these changes are yet unknown.

Tomlinson (1999) identified linkages within the globalization process with varying modalities ranging from "the social-institutional relationships between individuals and collectivities worldwide, the idea of the increasing 'flow' of goods, information, people and practices across national borders, and the more 'concrete' modalities of connection arising from technological developments such as the international system of rapid air transport and the more literal 'wiredness' of electronic communications systems".

Conceptual analysis on globalization often proceeds in a way which fails to substantially explore inequality in consumption patterns. This does not only fail to critically account for ecological breakdown rather ignores the interrelatedness and overlapping nature of societies as globalization presupposes the disappearance of boundaries (Oahme,1995 ;Gidens,1999).

Ohmae (1995) posits that the nation-state is becoming irrelevant with rise in capitalist market networks across national boundaries. A critical dimension of the resurgent capitalist market is the new consumption culture and preferences. Oahme (1995) identifies 'a cross-border civilization' and 'convergence of consumer tastes and preferences' with the emergence of 'global brands'. This aligns with "consumption identity" inscribed in the trademark of global corporations, which guides corporate and individual choices and consumption patterns. This has become a way of life for both individuals and groups.

Tomlison (1999) identified the increasing global connectivity and contends that the world is now a 'global unicity''-'a single social and cultural setting', stressing as Roland Robertson (1992) argued that 'globalization makes the world a 'single place'. This results in "homogenization of consumption" a conceptual exploration which underscores the increasing disappearance of variations in consumption patterns as a result of Western consumption hegemony linked to globalization.

In the particular case of the high income societies, Konya and Ohashi (2004) examined the increasing resurgence of homogeneous consumption patterns among the Organization for Economic Co-operation and Development (OECD) countries from 1985 to 1999. Their findings suggest that increased bilateral trade and foreign direct investment (FDI) has contributed to the convergence of consumption patterns.

There are related conceptual explorations which seek to create an interface between globalization and ecological security in the context of deleterious natural resource extraction. Rogers (1995) has provided such evidence which links ecology to globalization within security contexts, suggesting possible interface between "ecological security and multinational corporations". Rogers (1995) argued that ecological security refers to the creation of a condition where the physical surroundings of a community provide for the needs of its inhabitants without diminishing its natural stock.

Dalby (2002) has created a seminal linkage between environmental security and globalization, including human rights, ecological justice and gender inequality. Dalby (2013) argues that there is need to link environmental security to human survival. This has been part of the wider debates on 'redefining security from human dimension' (UNDP, 1994). The central thesis of the argument is that human security should be prioritized in the post -Cold War security debate against the traditional military/state centric notion of security which centered principally on protection of national boundaries and territorial integrity of the state was designed.

Appadurai (2003) re-echoes the new face of globalization whose "imagined world landscapes" reinforce the five dimensions globalization activities occur; *enthnoscapes*(people who move internationally),*technoscapes*(technology often linked to international corporations),*financescapes* (global capital, currency markets, stock exchanges),*mediascapes*(electronic and new media)and *ideoscapes*)official state ideologies and counter ideologies). The technological component is elucidated in Castell's (2000) "new society debate" in which he observed the rise in information technology and network hypertext.

In a similar line of thought, Beck (1992) had examined the corresponding effects of globalization and ecological breakdown resulting in the 'risk society' debate.

Despite the increasing ecological threats of globalization, proponents argue on 'technological solution' (Wackernagel, et al; 1999). This points to the debates of the ecological modernization theorists which argue that modern technological advancement could provide solution to environmental breakdown (Spaargaren & Mol, 1992). On the contrary, Schor (2005) argues that relying solely on technology would fail.

Our argument is linked to the later perspective as it examines the rationalities of globalization and preservation of the ecology where consumption has become a hegemony dominated and propagated by the West. This partly includes patterns of identity resulting in food consumption politics and inequality by the global food and beverage giants notably the McDonalds, Starbucks, Coca-Cola, Pepsi, Nestle etc. The vested interests of the industrialized West which exerts influence on ecological transformation of the poor societies, undermines resource renewal, greening and resource equity. Thus, conceptual debates advanced in this study suggest that globalization and the options for sustainable environmental consumption have been less emancipatory in the context of ecological justice and equal access to natural resource extraction.

GLOBALIZATION AND UNSUSTAINABLE ENVIRONMENTAL CONSUMPTION: MARXIAN POLITICAL ECOLOGY DEBATE

The Marxian political ecology analysis of globalization and unsustainable consumption provides patterns of inequality in resource extraction by the high-income societies (Bryant, 1998). Beyond this, ecological Marxism examines evidence of deleterious effects of resource exploitation in the Third World (Bryant, 1998; Amadi & Igwe, 2016).

In this context, and following the fact that no singular account both in ecological, economic, political or cultural perspectives have been accurately given on the sustainable consumption proclivity of globalization, the term remains rather contradictory. Rosenau (1996) reinforced this contradiction as he identified the complexities of globalization including widening the rich/poor gap. Thus, despite the premium placed on globalization by 'Western experts' a number of scholars hold contrary views. Stiglitz (2016) has been a vocal critique of globalization and reinforced "new discontent of globalization" an earlier debate in his 2003 book in which he argued that trade is among the major sources of discontent for a large share of Americans and Europe.

Globalization and environmental consumption is linked to the existential realities of the poor, since the lives of most rural poor largely derive from their natural environment (Amadi & Anokwuru, 2017). Thus, the understanding of ecological breakdown of the developing societies is important for international collaboration to redress the enormous challenges of globalization.

The perspectives on globalization within environmental consumption and sustainability contexts create dialectical interface with a number of interconnected indicators. The notion of interconnectivity of human activity with the natural environment to a large extent implies that globalization and environmental sustainability nexus could be complex. This interface as mediated by the Western capitalist interests is connected to the quest of the industrialized societies to gain economic and political control of the natural resources of the poor societies of the Third World. Therefore, globalization is not only an instrument to deepen hegemonic and economic interests but a multi-dimensional machination to deepen Western interest and overall ubiquity. Globalization and environmental sustainability nexus could be gleaned from a number of factors most notably ecological breakdown and patterns of unsustainable consumption. Most studies on the consumption patterns of the high-income societies point to unsustainability (Kove, 2003; Hobson, 2003; Davidson & Hatt, 2005; Collier, 2010; Amadi, Wordu & Ogbanga, 2015).

The understanding of the evidence of the ecological impacts of globalization could be examined from a number of studies on the measurement of the impact of globalization. The KOF index is one (Dreher, 2006; Dreher, Gaston & Martens, 2008). It uses three key dimensions defined as economic globalization, characterized as long distance flows of goods, capital and services as well as information and perceptions that accompany market exchanges; political globalization, characterized by a diffusion of government policies; and social globalization, expressed as the spread of ideas, information, images and people.

Seminal studies such as OECD (2013) examined the impact of globalization on the environment exploring origins and consequences. Niklas (2015) had provided useful evidence of the impact of globalization which is linked to complex interactions and effects to the society. The Globalization Indices (GI) provides country specific analysis of the impact of globalization on the basis of indicators. Such indices are evident in a number of seminal reports on the ecological impacts of consumption. In 1987, the Brundtland report argued that "An average person in North America consumes almost 20 times as much as a person in India or China, and 60 to 70 times more than a person in Bangladesh. It is simply impossible for the world as a whole to sustain a Western level of consumption for all. In fact, if 7 billion people were to consume as much energy and resources as well do in the west today we need 10 words, not one, to satisfy all our needs" (UN, 1987).

Related evidence is provided in validated indices that attempt to show evidence of ecological impacts of globalization such as the Maastricht Globalization Index (MGI) (Figge & Martens 2014; Martens & Raza 2009). Beyond these, Figge, et al. (2017) have conducted empirical exploration of the effects of globalization on ecological footprints. They found the divergent effects of globalization on ecological footprints.

Schor (2001) argued that "economic globalization, militarization, corruption, the monopolization of environmental resources, and the legacies of colonialism have meant that the global "South" doesn't consume enough—at least not in terms of basics such as food, clothing, shelter". Schor (2001) recounts that "in 1999, per capita GDP in the Less Developed Countries (or "global South") was $3,410 (measured as purchasing power parity). By contrast the Developed Countries ("global North") enjoyed average per capita GDP of $24,430, a gap of about eight times. One third of the population in the global South (1.2 billion) lives on less than $1 per day; 2.8 billion live on less than $2 per day. Together, this 4 billion comprises two-thirds of the world's population".

Thurow (1992) points out that "if the world 's population had the productivity of the Swiss, the consumption habits of the Chinese, the egalitarian instincts of the Swedes, and the social discipline of the Japanese, then the planet could support many times its current population without excessive pollution or deprivation for any-one".

The United Nations' Human Development Report (1998) states that; "runaway growth in consumption in the past fifty years is putting strains on the environment never before seen." In the United States, Schor, (2005) recounts that Americans consume at a higher rate than anywhere, and thus are the primary contributors to environmental issues caused by unsustainable consumption; issues such as forest destruction, ozone depletion, water and grain shortages, and soil loss. Over 100,000 synthetic chemicals are used in production, and almost none of these have been tested for their effect on the environment. U.S. industries admit to emitting over 4,000,000,000lbs of pollution a year and produce 22% of world's total industrial carbon dioxide emissions.

In the particular analysis on Australia, Hobson (2003) argued that the need to reduce consumption patterns of all social actors is critical to sustainable consumption.

She demonstrates how the consumption patterns of the high-income societies have been at variance with the principles of sustainable and equitable development. Schor (2001) critiqued the notion of "consumer culture" and suggests instead that consumption may significantly impact human welfare. In a distinct manner, Schor (2001) argued for instance, that, "New Coke is no better than Old Coke". Similar food giants that consume disproportionate resources include; Nestlé (Switzerland), PepsiCo (USA), Kraft (USA), ABinBev (Brazil), ADM (USA), Coca-Cola (USA), Mars Inc. (USA), Unilever (Netherlands), Tyson Foods (USA), Cargill (USA) (Berne Declaration, 2013).

A related study recounts that Nestlé is the world's largest food corporation with a turnover of US$ 103 billion. It has milk products, sweets, convenience foods, soft drinks, pet food, and health products which are sold in almost all countries of the world (Berne Declaration, 2013). Nestlé controls about 60% of the market for baby food in Latin America, and in Brazil up to 91% of the milk powder market. "Besides unethical advertising for baby food, Nestlé promotes its cereals through box-top tokens for free books for UK schools. Nestlé is criticized for using GM ingredients, for its purchasing policies for cocoa and coffee, for repression of trade unionists in Colombia, and for demanding an excessively high compensation payment for the nationalization of a Nestlé subsidiary during a famine in Ethiopia (Berne Declaration, 2013). In this regard, globalization allows the multinational corporations to plunder the globe's fragile ecosystem. In particular, as they pursue profits they produce consumer goods that ignore the true ecological implications.

Thus, a high emphasis was laid on globalization as a choice less purveyor of all round development. This forms part of the debate advanced by the ecological modernization theorists such as Gert Spaargaren who had accepted that sustainable consumption matters. Spaargaren (2003) contextualized consumption within the social practices model and argued that the greening of social practices matters.

Evidence of global concern on the challenges of unsustainable consumption deepens. For instance, in 2002 at the Johannesburg Plan of Implementation, the UN (2002) reported that; 'The global environment continues to suffer. Loss of biodiversity continues, fish stocks continue to be depleted, desertification and loss of fertile land". This has reopened debates on the global resource politics and natural resource wars which have provided important insights on the problems of globalization and unsustainable consumption.

Related account documents competition among the global multinationals to control "abundant resources" in the South, resulting in resource conflicts. This contention in the 2000s was popularized by the International Peace Research Institute, Oslo (PRIO), which posits that environmental security threats and conflicts in the global South is an attribute of violence over the struggle to control abundant resources (de Soyza, 2000).This debunked the neo Malthusian conceptualizations of the Toronto

Table 2. Technological innovation with significant environmental impact

Technological Innovation with Significant Environmental Impact

Innovation	Form of Change	Primary Investment	Secondary Environmental Effect	Primary Motivation
Coal scrubbers	End-of-pipe	Reduced SO2 emissions	Increased energy use (-)	Environmental
Electric arc furnace	Process Reduced	Energy consumption Increased	use of scrap (+/-)	Economic
HCFCs	Input substitution	Reduced ozone depletion		Environmental
Biodegradable packaging	Product change	Reduced waste accumulation	Reduced waste from plastics manufacturing (+)	Environmental
Thermo-mech. pulping	Process	Reduced waste water discharges	Increased energy use (-)	Economic
Low-solvent paint	Product change	Reduced smog		Environmental
Reverse osmosis purification	End-of-pipe	Reduced waste water discharges	Increased solid waste (-)	Environmental
Counter-current rinsing	Process	Reduced heavy metal waste	Reduced metal inputs (+)	Environmental/ Economic

Source: Johnstone, N. "Globalization, Technology, and Environment" OECD, 1997

School and their scarcity debate which posits that population growth results in competition over scare natural resources in the poor societies, which in turn results in violent conflicts (Homer-Dixon,1991).

Similarly, Harvey (2005) re-echoes the persistent deleterious natural resource use as a dominant factor in the neo liberal order. For instance, increasing environmental pollution arising from the multinational oil corporations (MNOCs) and oil resource extraction in the poor societies such as the Niger Delta in Nigeria has been at issue in sustainability debate. UNEP (2011) reports that it will take 25 to 30 years for an effective clean- up of the oil resource induced pollution of Ogoniland a community in the Niger Delta, Nigeria.

OECD, (2013:113) argues that "developed countries – the pioneers of global industrialization – were the world's biggest polluters, responsible for the lion's share of GHG emissions. Today, the United States is responsible for around 20% of global GHG emissions". OECD (2013:112) re-echoes that globalization promotes CO_2 emissions from transport both road transports (including cars and lories) and air transport. OECD (2013) reports that "aviation is today responsible for 4-9% of total GHG emissions released into the atmosphere. Between 1990 and 2004, GHG emissions from aviation increased to 86%".

The persistent amazon rain forest deforestation in Brazil (Ometto, Aguiar & Martinelli, 2014) and natural forest deforestation in Nigeria (Butler, 2005), reveal the distortion of wildlife and the ecosystem. Much of the deforestation is linked to global multinationals and resource exploration. This points out the contradictions of the sustainability paradigm as the flora and fauna, plant and wild life species are decimated. Also in the background are increased regional tensions arising from cultural homogenization in which consumption culture a Western hegemony is spreading through globalization among in the poor societies (Davidson & Hatt, 2005; Amadi, Igwe & Ogbanga, 2016).

Klare (1996) recounts that the environment is endangered and argued that 'the uneven impact of global environmental decline is seen in many areas and that the first to suffer are those in marginally habitable area'. He points out that the last 20 years has seen the destruction of coastal marine habitats, increases in coastal pollution and a shrinking of the marine fish catch.

There is also the assumption that by creating eco-friendly environment equitable development could be achieved (Hawken, Lovins & Lovins, 1999; Kovel, 2000). The ecological footprint debate presupposes the necessary proportionate land area that can sustain consumption. A key argument from the "ecological footprint" perspective is that as urban areas grow and develop their reliance on the natural resources to keep up with their development intensifies in order to meet production and consumption demands. This has deleterious effects on the natural environment as it is persistently exploited (Chambers, Simmons, & Wackernagel, 2001). Thus, various levels of sustainability spaces for development and policy framings are open for further discourse

RECOMMENDATIONS

The chapter seeks for alternative approach in which sustained policy framings and international development collaboration could be institutionalized to strengthen sustainable environmental consumption. This could be premised on ecological justice and natural resource equity. The new policy engagement includes international collaboration and cooperation for environmental sustainability. One which should be change oriented and transform international relations from resource competition to equitable and transparent global resource collaboration among the rich and poor societies. The role of the international community should be to mediate and reconcile the environment with sustainability and equally to recreate the understanding of consumption, equality and remediation of environmental degradation and food insecurity.

Globalization as the chapter demonstrates divides environmental sustainability into asymmetrical and deleterious domains. This poses challenges to the poor in various ways. Global policy intervention and response to unsustainable environmental consumption is suggested. The Rio +20 Report, emphasizes that green economy should contribute to "eradicating poverty as well as sustained economic growth, enhancing social inclusion, improving human welfare and creating opportunities for employment and decent work for all, while maintaining the healthy functioning of the Earth's ecosystems" (Rio Report, 2012).

Strengthening environmental sustainability through traditional ecological knowledge (TEK) is important. This involves the integration of traditional and modern ecological knowledge to address ecological problems. Such collaboration at individual and cross country levels is expedient. This should be a central global policy priority to mitigate inequality, poverty, discrimination and deprivation of the developing countries partly emanating from the deleterious resource consumption patterns of the affluent societies.

There is equally increasing need to evolve indigenous strategies of resource management and related approaches to sustainable environmental consumption. Green urbanization constructs -the propagation of eco communities and in particular, the need to address resurgent challenges of climate change vulnerability, sustainable urban ecology sprawl, resource renewal, equity and social inclusion in cities are worthwhile.

A correlate of sustainable consumption is sustainable production. Thus, social actors and global corporate production giants should device collaborative strategy of sustainable production one which is not largely informed by profit motives but ecological considerations in which the ecosystem should not be tainted. Thus, there should be building of resilience blocs to unsustainable production and consumption through collaboration, networks and movements.

There is need for a change in lifestyle and consumption patterns of the affluent societies. Pro-environment and inclusive consumption pattern is suggested to mitigate inequality and in particular, to deploy a bottom- top approach to consumption. This takes account of the poor and their consumption needs which remains subsumed by Western interests. In particular, the chapter calls for a new— international pressure in promoting political ecology aimed at institutionalizing equality in natural resource extraction, resource transparency, efficiency and ecological justice. The aim is to institutionalize a green consumption which preserves the ecosystem as it meets the consumption needs of the consumer and protects the natural environment.

To recreate unsustainable consumption, a number of green movements had emerged across Europe and North America propagating such famous dictum as; *'Think globally, act locally'* which according to Tomlinson (1995) suggests a political strategy motivated by a very clear collective cultural narrative of what the 'good life' entails. Putting these trends in action within global policy response and collaboration could be helpful.

FUTURE RESEARCH DIRECTIONS

Beyond the environmental component which includes the ecosystem, natural resource extraction and consumption, globalization needs to go into new directions if it is to make a contribution in every aspect of sustainable development. A central research direction of globalization and sustainable consumption dynamics is to understand the 'phenomenology' of ecological challenges of unsustainable consumption mediated by capitalist resource appropriation. Sustainable consumption aims at poverty eradication, ecological justice, social protection, inclusion and equality including resource consciousness and values. It entails sustainable use of natural resources such as water, crude oil resource extraction, gold, diamond, etc. It extends to issues of energy consumption, clean energy and global resource governance including green energy consumption, resource transformation and decoupling, de-materialism and ecological footprints through green growth and low carbon transition.

It is less difficult to see that the horizon of significance made available by globalization suggests possibilities not only for the reconstitution of the traditional uses and depletion of the natural environment but also for novel research agenda that could examine new terminological issues associated with sustainability. This suggests a new research agenda on globalization and natural resource consumption such as the moral economy, ethics of sustainability, green consumption, ethical globalization, natural resource accounting, ecological justice etc. Thus, the workings of globalization in the context of sustainable consumption remain contestable in terms of ecological justice and equity. This requires further critical study.

CONCLUSION

The point the chapter has been making is that global environment is being destroyed by processes that exist primarily to fuel the North's need to constantly consume. This as the chapter suggests at intervals is linked to globalization. Although globalization may not be entirely antithetical to the environment, our emphasis has been on how to reduce deleterious consumption, improve unequal access and resource extraction.

Our thesis has been that the advanced societies deploy globalization apparatus to the detriment of the poor. Globalization as the chapter repeatedly emphasizes burdens much of the South with unequal environmental costs and low environmental standards to the advantage of the North. It drives over-consumption by the societies of the North resulting in resource distortions and consumption imbalances in the South. The chapter demonstrates that this puts global consumption beyond earth's carrying capacity. This equally results in ecological breakdown which affects human and non- human species in the ecosystem.

The chapter has demonstrated that environmental sustainability aims at rethinking some of the foundational questions of deleterious environmental consumption. In a distinct manner, it has demonstrated that in spite of the less clear linkages between globalization and sustainable consumption, a number of evidence points out deleterious consumption in ways which put the domain of sustainable consumption firmly on global policy framing.

In particular, sustainability in ecological contexts should challenge and redirect the resource inequality and ecological injustice in which the affluent societies disproportionately appropriate the natural resources. This as the chapter argues at intervals results in environmental degradation such as pollution, deforestation, acid rains, gas flaring, global warming etc linked to globalization.

The fact that individual actions are intimately connected with consumption, suggests that globalization is largely linked to consumption. This includes the network of food chains, electronic networks, global water corporations etc. Thus, environmental consumption system is determined by large global interconnected structures which involve global and local actors. Consumption politics remains at the center of this interconnection. For instance, the lifestyles and consumption patterns of the affluent societies including the golf courses etc have been elements of asymmetrical resource use. This often has dire ecological consequences. A key strand of the persistent unsustainable consumption is patterns of lifestyle of the high income societies of the North. This drives consumption patterns, choices and preferences often linked to addiction, use of luxury goods, perhaps at the expense of the environment.

The chapter suggests that the sustainability conception of globalization appears to be at variance with the logic of capitalism, which fosters inequality, the dominance of the capitalist exploitation and the global interest of the international system. This negatively influences ecological choices in various ways. This work has offered a means of taking seriously the global power structure which undermines resourcefulness and ecological accountability. The work insists on equitable and renewable resource consumption across rich and poor societies. This is of immense value to humanity and the environment and cuts across disciplinary norms and in particular, aims to recreate the capitalist mode of natural resource exploitation, suggesting the critical need to foster equitable relationship between globalization and sustainable consumption.

REFERENCES

Amadi, L & Agena, J .(2015).Globalization Culture Mutation and New Identity. *African Journal of Culture and History*, 7(1), 16-27.

Amadi, L., & Anokwuru, G. (2017). Sustainable Rural Livelihoods: Elusive Post-Colonial Development Project in Nigeria? *International Journal of Poultry Science*, 2(4), 1–16.

Amadi, L., & Igwe, P. (2016). Maximizing the Eco Tourism Potentials of the Wetland Regions through Sustainable Environmental Consumption: A Case of the Niger Delta, Nigeria. *The Journal of Social Sciences Research*, 2(1), 13–22.

Amadi, L., Igwe, P., & Ogbanga, M. (2016). Talking Right, Walking Wrong: Global Environmental Negotiations and Unsustainable Environmental Consumption. *International Journal of Research in Environmental Science*, 2(2), 24–38.

Amadi, L., & Imoh-ita, I. (2017). Intellectual capital and environmental sustainability measurement nexus: A review of the literature *Int. J. Learning and Intellectual Capital*, X14(2), 154–176. doi:10.1504/IJLIC.2017.084071

Amadi, L., Wordu, S., & Ogbanga, M. (2015). Sustainable Development in Crisis? A Post Development Perspective. *Journal of Sustainable Development in Africa*, 17(1), 140–163.

Appadurai, A. (1996). *Modernity at Large: Cultural Dimensions of Globalization*. Minneapolis, MN: University of Minnesota Press.

Appadurai, A. (2003). Modernity at Large: Cultural Dimensions of Globalization, *The Russian. The Sociological Review*, 3(4), 57–66.

Bakan, J. (2004). *The Corporation: The Pathological Pursuit of Profit & Power*. Toronto: Penguin Group.

Bauman, Z. (1998). *Work, Consumerism and the New Poor*. Philadelphia: Open University Press.

Beck, U. (1992). *Risk Society. Towards a New Modernity*. London: Sage.

Beitz, C. (2001). Does Global Inequality Matter? In T. W. Pogge (Ed.), *Global Justice*. Oxford. doi:10.1111/1467-9973.00177

Bello, W. (2003). *De-globalization Ideas for a New World Economy*. Fernwood Publishing Limited.

Blaikie, P., & Brookfield, H. (1987). *Land Degradation and Society*. London: Methuen.

Bryant, R. (1998). Power, Knowledge and Political Ecology. *Progress in Physical Geography*, *22*(1), 79–94. doi:10.1177/030913339802200104

Butler, R. (2005). *Nigeria has worst deforestation rate, FAO revises*. Available at https://news.mongabay.com/2005/11/nigeria-has-worst-deforestation-rate-fao revises-figures/Accesses 18/4/2017

Callinicos, A. (2009). *Imperialism and Global Political Economy*. London: Polity Press.

Carson, R. (1962). *Silent Spring*. New York: Houghton Mifflin.

Castells, M. (2000a). The Information Age: Economy: Society, and Culture (2nd ed.; Vols. 1-3). Maiden, MA: Blackwell.

Chambers, N., Simmons, C., & Wackernagel, M. (2001). *Sharing nature's interest: Ecological footprints as an indicator of sustainability*. London: Earthscan Publications.

Chow, L. (2017). *EPA Chief Denies CO2 as Primary Driver of Climate Change*. Eco Watch.

Collier, P. (2010). *The Plundered Planet. Why We Must--and How We Can--Manage Nature for Global Prosperity*. Oxford Press.

Collier, P., & Gunning, J. (1999). Why Has Africa Grown Slowly? *The Journal of Economic Perspectives*, *13*(3), 3–22. doi:10.1257/jep.13.3.3

Dalby, S. (2002). *Environmental Security*. Minneapolis, MN: University of Minosota Press.

Dalby, S. (2013). Environmental dimensions of human security. In R. Floyd & R. Mathew (Eds.), *Environmental Security Approaches and Issues*. London Routledge Taylor and Francis.

Davidson, D., & Hatt, K. (2005). *Consuming Sustainability Critical Social Analysis of Ecological Change*. Fernwood Publishing.

de Soyza, I. (2000). The resource curse: Are civil wars driven by rapacity or paucity? In M. Berdal & D. M. Malone (Eds.), *Greed and grievance: Economic agendas in civil wars* (pp. 113–135). Boulder, CO: Lynne Rienner.

Dreher, A. (2006). Does globalization affect growth? Evidence from a new index of globalization. *Applied Economics*, *38*(10), 1091–1110. doi:10.1080/00036840500392078

Dreher, A., Gaston, N., & Martens, P. (2008). *Measuring globalisation: Gauging its consequences*. New York: Springer. doi:10.1007/978-0-387-74069-0

Driessen, P. (2003). *Eco-Imperialism: Green Power, Black Death*. Belleview, WA: Free Enterprise Press.

Esty, D., & Pangestu, M. (1999). *Globalization and The Environment in Asia*. United States-Asia Environmental Partnership Framing Paper.

Fayiga, A., Ipinmoroti, M., & Chirenje, T. (2017). Environmental pollution in Africa. *Environment, Development and Sustainability*. doi:10.100710668-016-9894-4

Ferguson, J. (2006). *Decomposing Modernity: History and Hierarchy after Development*. Irvine, CA: Department of Anthropology University of California. doi:10.1215/9780822387640-008

Figge, F., & Martens, P. (2014). *Globalization Continues: The Maastricht Globalisation Index Revisited and Updated Globalizations*. Academic Press. 10.1080/14747731.2014.887389

Figge, L., Oebels, K., & Offermans, A. (2017). The effects of globalization on Ecological Footprints:an empirical analysis. *Environment, Development and Sustainability*, *19*(3), 863–876. doi:10.100710668-016-9769-8

Giddens, A. (1999). *Runaway World: How Globalization is Reshaping our Lives*. London: Profile.

Goodland, R. (1995). The Concept of Environmental Sustainability. *Annual Review of Ecology and Systematics*, *26*(1), 1–24. doi:10.1146/annurev.es.26.110195.000245

Haley, E., Hatt, K., & Tunstall, R. (2005). *You are What You eat. In Consuming Sustainability Critical Social Analysis of Ecological Change*. Fernwood Publishing.

Harvey, D. (2005). *Brief History of Neoliberalism*. New York: Oxford University Press.

Hawken, P., Lovins, A., & Lovins, L. (1999). *Natural capitalism: creating the next industrial revolution*. Boston: Little, Brown and Company.

Hobson, K. (2003). Consumption, Environmental Sustainability and Human Geography in Australia: A Missing Research Agenda? *Australian Geographical Studies*, *41*(2), 148–155. doi:10.1111/1467-8470.00201

Hoekstra, A., & Wiedmann, T. (2014). Humanity's unsustainable environmental footprint. *Science*, *344*(6188), 1114-1117. DOI: 10.1126 science.1248365

Homer-Dixon, T. (1991). On the Threshold: Environmental Changes as Causes of Acute Conflict. *International Security*, *16*(2), 76–116. doi:10.2307/2539061

Jorgenson, A., & Kick, E. (2003). Globalization and the environment. *Journal of World-systems Research, 9*(2), 195–205. doi:10.5195/JWSR.2003.243

Klare, M. (1996, November). Redefining Security: The New Global Schisms. *Current History (New York, N.Y.)*.

Konya, I., & Ohashiz, H. (2004). *Globalization and Consumption Patterns among the OECD Countries*. Boston College Working Papers in Economics.

Korten, D. (1995). *When Corporations Rule the World*. Kumarian Press Inc./Berrett-Koehler.

Kovel, J. (2000). The Struggle for Use Value. *Capitalism, Nature, Socialism, 11*(2), 3–23. doi:10.1080/10455750009358910

Kristof, N. (1997, January 9). For Third World water is still a deadly drink. *New York Times*, p. 2.

Lockwood, B., & Redoano, M. (2005). *The CSGR Globalization Index: An Introductory Guide Centre for the Study of Globalization and Regionalization working paper 155/04*. Retrieved from www2.warwick,ac.uk.fac/soc.csgr/index/citation

Lothar, B. (1991). Peace Through Parks: The Environment on the Peace Research Agenda. *Journal of Peace Research, 28*(40), 407–423.

Martens, P., Akin, S., Maud, H., & Mohsin, R. (2010). Is globalization healthy: A statistical indicator analysis of the impacts of globalization on health. *Globalization and Health, 6*(16). doi:10.1186/1744-8603-6-16 PMID:20849605

Martens, P., & Raza, M. (2009). Globalization in the 21st century: Measuring regional changes in multiple domains. *Integrated Assessment, 9*(1), 1–18.

Martens, P., & Raza, M. (2010). Is globalization sustainable? *Sustainability, 2*(1), 280–293. doi:10.3390u2010280

McGrew, A. (1992). *A Global Society?* London: Polity Press.

Meadows, D. (1972). *The Limits to Growth: A report for the Club of Rome's project on the predicament of mankind, Part 1 Club of Rome*. Potomac Associates.

Miller, M. (1995). *The Third World in global environmental politics*. Boulder, CO: Lynne Rienner Publishers.

Niklas, P. (2015). The Evidence on Globalization. *World Economy, 38*(3), 509–552. doi:10.1111/twec.12174

OECD. (2008). *Handbook on constructing composite indicators: Methodology and user guide*. Paris: OECD Publishing.

OECD. (2013). What is the impact of globalization on the environment? In *Economic Globalization: Origins and consequences*. Paris: OECD Publishing; doi:10.1787/9789264111905-8-

Ohmae, K. (1995). *The End of the Nation State*. New York: Free Press.

Ometto, J., Aguiar, A., & Martinelli, L. (2014). Amazon deforestation in Brazil: Effects, drivers and challenges. *Carbon Management*, *2*(5), 575–585. doi:10.4155/cmt.11.48

Rennen, W., & Martens, P. (2003). The globalization timeline. *Integrated Assessment*, *4*(3), 137–144. doi:10.1076/iaij.4.3.137.23768

Rio Report, U. N. (2012). *The Future We Want*. Oxford Press.

Roberts, J., & Thanos, N. (Eds.). (2003). *Trouble in Paradise: Globalization and Environmental Crises in Latin America*. New York: Routledge.

Robertson, R. (1992). *Globalization: Social Theory and Global Culture*. Sage.

Rogers, K. (1997). *Ecological Security and Multinational Corporations*. Kluwer Academic Publishers.

Rosenau, J. (1996). *Complexities and Contradictions of Globalization. World Politics*. Cambridge, UK: Cambridge University Press.

Scheren, P., Ibe, A., Janssen, F., & Lemmens, A. (2002). Environmental pollution in the Gulf of Guinea – a regional approach. *Marine Pollution Bulletin*, *44*(7), 633–641. doi:10.1016/S0025-326X(01)00305-8 PMID:12222886

Schor, J. (2001). *Why Do We Consume So Much?* Saint John's University Clemens Lecture Series.

Schor, J. (2005). Prices and quantities: Unsustainable consumption and the global economy. *Ecological Economics*, *55*(3), 309–320. doi:10.1016/j.ecolecon.2005.07.030

Shiva, V. (1988). *Staying alive: Women, ecology, and development*. London: Zed Books.

Shove, E. (2003). Converging Conventions of Comfort, Cleanliness and Convenience. *Journal of Consumer Policy*, *26*(4), 395–418. doi:10.1023/A:1026362829781

Smith, J. (2017). How are GMOs and Roundup linked to cancer? *International Conference on The Truth About Cancer Orlando United States*.

Spaargaren, G. (2003). Sustainable Consumption: A Theoretical and Environmental Policy Perspective. *Society & Natural Resources*, *16*(8), 687–701. doi:10.1080/08941920309192

Spaargaren, G., & Mol, A. (1992). Sociology, Environment and Modernity: Ecological Modernization as a Theory of Social Change. *Society & Natural Resources*, •••, 5.

Speake, S., & Gismondi, M. (2005). Water: A Human Right. In *Consuming Sustainability Critical Social Analysis of Ecological Change*. Fernwood Publishing.

Stern, P. (1997). *Environmentally Significant Consumption. Research Directions*. National Academy Press.

Stiglitz, J. (2003). *Globalization and Its Discontents*. London: Cambridge University Press.

Stiglitz, J. (2016). *Globalization and its New Discontents*. Project Syndicate.

Terhune, C. (2004, June 18). Coke CEO Says obesity is a challenge. *Wall Street Journal*.

The Berne Declaration. (2013). *Agropolicy. A handful of corporations control world food production*. Available at http://www.econexus.info/sites/econexus/files/Agropoly_Econexus_BerneDeclaration

Thurow, L. (1992). *The Coming Economic Battle Among Japan America, and Europe*. New York: William Marrow and Company Inc.

Tomlinson, J. (1999). *Globalization and Culture*. University of Chicago Press.

UN. (2002a). *Johannesburg declaration on sustainable development*. United Nations.

UNDP - Human Development Report. (1998). *Brazil -Globalization and Changes in Consumer Patterns*. Oxford Press.

UNEP (United Nations Environment Programme). (2011). Environmental Assessment of Ogoniland. UNEP Report.

United Nations. (1987). *Our Common Future. Report of the World Commission on Environment and Development*. Oxford Press.

United Nations (UN). (1997). Commission on Sustainable Development. Comprehensive assessment of the freshwater resources of the world. New York: UN.

United Nations Development Programme. (1994). *Human Development Report: Redefining Security*. Oxford, UK: Human Dimensions.

Veseth, M. (2004). *What is International Political Economy?* UNESCO.

Vitousek, P., Ehrlich, P., Ehrlich, A., & Matson, P. (1986). Human appropriation of the products of photosynthesis. *Bioscience*, *36*(6), 368–373. doi:10.2307/1310258

Wackernagel, M., Onisto, L., Bello, P., Callejas Linares, A., Susana López Falfán, I., Méndez García, J., ... Guadalupe Suárez Guerrero, M. (1999). National natural capital accounting with the ecological footprint concept. *Ecological Economics*, *29*(3), 375–390. doi:10.1016/S0921-8009(98)90063-5

White, L. (1967). The historical roots of our ecological crisis. *Science*, *155*(3767), 1203–1207. doi:10.1126cience.155.3767.1203 PMID:17847526

Wise, T. (2015). Two Roads Diverged in the Food Crisis: Global policy takes the one moreq traveled. *Canadian Food Studies*, *2*(2), 1-15.

World Bank. (2015). World Development Indicators. Washington, DC: World Bank.

World Health Organization. (1997). Health and environment in sustainable development: Five years after the earth summit. Geneva: World Health Organization. (No. WHO/EHG/97.8)

Worzel, R. (1994). *Facing the Future; The Seven Forces Revolutionalizing Our Lives*. Stoddart Publishing.

KEY TERMS AND DEFINITIONS

Ecological Footprint: The understanding of the environmental impact of an individual on the ecosystem, measured by a wide range of goods consumed on a daily basis or in a given period of time. It includes the area of productive land and water required on a continuous basis to produce the resources consumed and to assimilate the wastes produced by a defined population wherever on earth that land is located.

Ecological Justice: A notion that suggests that the ecosystem should be treated fairly with some level of justice.

Ecological Modernization: A debate that ecological reforms through technological advancement could provide solutions to environmental degradation and similar challenges.

Green Consumerism: A consumption pattern that recognizes the natural environment or the ecosystem.

Political Ecology: An aspect of development studies that examines the politicization of nature and inequality in access, use, and control of the natural environment as well as its effects on human beings and the ecosystem.

Chapter 2
From Kyoto to Paris:
An Analysis of the Politics of Multilateralism on Climate Change

Moses Metumara Duruji
Covenant University, Nigeria

Faith O. Olanrewaju
Covenant University, Nigeria

Favour U. Duruji-Moses
Covenant University, Nigeria

ABSTRACT

The Earth Summit of 1992 held in Rio de Janeiro awakened the consciousness of the world to the danger of climate change. The establishment of the United Nations Framework Convention on Climate Change provided the platform for parties to negotiate on ways of moving forward. The global acknowledgement of the weightiness of the climate change and the future of the planet galvanized international agreements to this regard. Consequently, a landmark agreement was brokered in 1992 at Kyoto, Japan and 2015 in Paris, France. However, the strong issues of national interest tend to bedevil the implementation that would take the world forward on climate change. The chapter therefore examined multilateralism from the platform of climate change conferences and analyzed the political undertone behind disappointing outcomes even when most of the negotiators realized that the only way to salvage the impending doom is a multilateral binding agreement when nation-state can subsume their narrow interest.

DOI: 10.4018/978-1-5225-3990-2.ch002

INTRODUCTION

Climate change instigated by anthropogenic greenhouse gases has arisen as one of the most significant environmental concerns confronting the international community. For example, greenhouse gases-especially fossil fuel-based carbon dioxide emissions gathers in the atmosphere as an outcome of human activities. These progressive intensifications in greenhouse gas concentrations cause various changes in the climates such as a rise in the world's average temperature (Bohringer, 2003). Since the end of the last century and the beginning of the 21st century, its impacts on weather and other natural environmental heritages have become increasingly felt.

As such, climate change is now globally acknowledged as one of the most important challenges facing the world. Its impacts on the society and environment are unprecedented and better imagined that real. Nonetheless scholars of climate change have recommended that global greenhouse gas emissions must reduce swiftly to mitigate the impending consequences resulting from increase in temperature. This in the actual sense is the scientific underpinning of this political problem (Shanahan, 2009).

Therefore, the dire need for humanity to find solution to these problems have created an awareness amongst nations of the world to embrace a collective mechanism or multilateral approach through a platform called United Nations Framework Convention on Climate Change (UNFCCC) established under the auspices of the United Nations.

The paper therefore examined multilateralism from the platform of climate change conferences and analyzed the political undertone behind disappointing outcomes even when most of the negotiators realized that the only way to salvage the impending doom is a multilateral binding agreement.

METHODOLOGY

The paper relied heavily on secondary sources especially commentaries and reports arising from the UNFCCC Conference of the Parties (COP) conferences. Books, newspaper reports, conference materials and materials sourced from the internet were most useful. Data sourced through these were analyzed through the employment of qualitative descriptive analysis with the backdrop of theory of multilateralism.

THE CONCEPT OF MULTILATERALISM

Multilateralism is a concept of international relations that is different from unilateralism, bilateralism or regionalism. It is 'the practice of coordinating national policies in groups of three or more states, through *ad hoc* arrangements or by means of an entrenched institutions' (Keohane, 1990, p.731; Yarbrough and Yarbrough, 1992). To Ruggie (1992, pp.567-568), multilateralism meant 'coordinating relations among three or more states', but 'in accordance with certain principles' that govern dealings between them. These definitions limit it to arrangements involving states. It focuses mainly (albeit not exclusively) on institutions, defined as 'inherited patterns of rules and relationships that can affect beliefs and expectations, and as potential tools for the pursuit of their own objectives' (Keohane, 2000 p. 96; Keohane and Nye, 2000a; 2000b).

But multilateral cooperation occurs between states as well as other concerned global stakeholders to arrive at a solution to common concerns. Multilateralism is voluntary and (more or less) institutionalized cooperation governed by principles and norms, with rules that apply (more or less) equally to all. Inherent in the conceptualization of multilateralism is the concept of diffuse reciprocity which was first discussed by Keohane (1986). It means that state actors expect to achieve gains from multilateralism in the long term and on a variety of issues. In other words, they expect the arrangement to 'yield a rough equivalence of benefits in the aggregate and over time' (Ruggie, 1992 p.571). Multilateralism becomes institutionalized when 'multilateral arrangements with persistent rules' emerge (Keohane, 1990 p. 733).

In modern times, multilateral agreements sprung up mainly to manage relations amongst states based on the principle of state sovereignty. Contrary to this, in the 17th century, multilateral arrangements manage property issues, such as the control of oceans. However, until the 19th century multilateral cooperation, was relatively rare. This century witnessed more frequent multilateral cooperation with the signature of several treaties on issues including trade, river transport and public health.

One effect of growing interdependence was to internationalize issues once considered to be strictly national. Most multilateral agreements during the 19th century, however, did not lead to the creation of formal organizations. In contrast to previous forms, the 20th century multilateralism brought about the formation of formal multilateral organizations, including a multi-purpose organization with universal membership. A shift from loose, informal agreements to formal organizations inevitably had an impact on the International Order (Ruggie, 1992). Conversely, common standards and regulations were developed to facilitate economic exchanges. The economies of major powers also became increasingly interdependent, thus encouraging the recognition of common interests (Armstrong and Lloyd et. al., 2004).

It is based on this premise that the UNFCCC was created to provide mechanism for stakeholders to formulate a mutual strategic understanding or view on mutual climate change goals and to institute collaborative activities that would offer common solutions to the global problem of climate change. It also provides a platform for negotiation on climate change policies, exchange of opinions on salient issues in climate change debates and development of concrete activities to curb climate change through the carrying out of specific cooperative tasks (Romano, 2010).

The gathering of Kyoto Japan in 1997 fulfilled aspects of these issues in the sense that an agreement considered significant to achieve the objective of the gathering emerged but owing to attitude of state actors in its follow up weakened the contents of the document. Subsequently the inability to achieve progress since 2005 when Russia ratified the protocol to the conference in Copenhagen and Cancun tend to give credence to alternative views of multilateralism. For instance, Martin (1992) acknowledges that multilateralism at times may not be the most effective strategy of promoting international cooperation especially when governments have doubts about the repercussion of 'losing today' without considering the long term mutually beneficial effect. Obviously companionable with a view of multilateralism as an anachronism is one that considers it a 'weapon of the weak' (Kagan, 2002 p.4). Put differently, states that pursue multilateral agreements are those that lack the power, however measured, to enforce solutions to international problems that are of interests to them (Caroline & Peterson 2009).

THE UNITED NATIONS FRAMEWORK CONVENTION ON CLIMATE CHANGE

The UN climate-change negotiations often take place under the UN Framework Convention on Climate Change (UNFCCC), which is an international treaty formed at the Earth Summit in Rio de Janeiro, Brazil, in 1992 to avert hazardous climate change ensuing from emissions of greenhouse gases. A total of 192 Parties ratified the UNFCCC. It entered into force in 1994. Under the auspices of the Convention, countries consented to protecting the climate system for both the present and future generations according to their 'common but differentiated responsibilities and respective capabilities', meaning that developed countries 'should take the lead in combating climate change and its adverse effects'. Parties also agreed that the degree to which developing countries can meet their treaty obligations is dependent on the extent to which developed countries provide technology and finance. It was also established that 'economic and social development and poverty eradication are the first and overriding priorities of the developing country parties' (Shanahan, 2009).

Therefore, the UNFCCC's stated goal is the "stabilization of greenhouse gas concentrations in the atmosphere at a level that would prevent dangerous anthropogenic interference with the climate system" (UNFCCC, 1992, Article 2).

SUSTAINABLE DEVELOPMENT AND CLIMATE CHANGE

The United Nations Conference on Environment and Development, the earth summit that took place in Rio de Janeiro Brazil was the largest ever international conference and the central aim was to identify the principles of action towards "sustainable development" in the future (Elliot, 1999, p. 4). According to Adam, (1990), the term sustainable development has gained ground beyond the confines of global environmental organization as it has become embraced in the political and academic field. According to Mather and Chapman (1995) the primary output of the Agenda 21 document of UN Conference on Environment and Development was driven at reconciling conservation actions into the 21st century. More so, it contained a substantial debate over the meaning and practice of sustainable development.

Literarily, sustainable development refers to maintaining development over time. Turner, (1988 p.12) defines it as follows;

In principle, such an optimal (sustainable growth) policy would seek to maintain an "acceptable" rate of growth in per-capital real incomes without depleting the national capital asset stock or the natural environment asset stock. Development that meets the needs of the present without compromising the ability of future generations to meet their own needs. (World Commission on Environment and Development, 1987 p.43)

It can be correctly stated that sustainable development is fundamentally about reconciling development and the environmental resources on which the society depends (Elliot, 1999).

The challenges of sustainable development include clearing off the contamination or pollution impacts of the past development. Most times the impacts of pollutions do not occur immediately as is the case of climate change. It accumulates over time. This means that the need for sustainable development in the future is also confirmed by the human cost of patterns and processes of development to date (Elliot, 1999).

The Kyoto Japan Conference

The Kyoto Protocol was negotiated during the third conference of the parties (COP3) to the United Nations Framework Convention of Climate Change. The Conference was held between 1 and 11 December 1997 in Kyoto, Japan. The detailed rules for the implementation of the Protocol called the "Marrakech Accords" were adopted at COP 7 in Marrakech in 2001. The Conference was attended by over 10,000 participants, covering delegates from International Governmental Organizations (IGOs), governments and Non-Governmental Organizations (NGOs) and the press. The conference included high-level section which featured statements from over 125 ministers. After intense informal and formal negotiations spanning week and a half, Parties to the UNFCCC adopted the Kyoto Protocol on 11 December 1997 (IISD, 2007).

The Climate Change Convention provided the institutional framework for international climate policy. Most of the countries that are Parties to the agreement have ratified the Convention. Parties to the Climate Change Convention hold periodic meetings called Conferences of Parties (COP) aimed at promoting and reviewing efforts to contend with global warming.

Kyoto Protocol and Issues Arising From It

In 1997, Parties to the UNFCCC adopted a document tagged Kyoto Protocol which outlines agreements reached at the Conference. This agreement which is meant to be binding on all parties on ratification by a given number of them came into force in 2005 after Russia ratified. Ever since, over 189 countries have ratified the protocol. Apart from the fact that it sets targets for developed countries, it was also legally binding, and created vital international monitoring, reporting and verification instruments to ensure compliance. Developed countries were obliged to limit their emissions to an average of 5.2 per cent below 1990 levels between 2008 and 2012. To ensure this is achieved, the protocol created 'flexibility mechanisms' – such as carbon trading and Clean Development Mechanism (CDM), which permits developed countries to achieve their target emission goals by financing emissions reductions in developing countries (Shanahan, 2009; IISD, 1997).

The key characteristics of the Kyoto Protocol is that it established obligatory/mandatory targets for the European community and 37 industrialized countries for reducing greenhouse gas (GHG) emissions, amounting to an average of 5 percent against 1990 levels over the five-year period 2008-2012 (Friends of the Earth, 2009). It necessitates industrialized countries as listed in its Annex B - to reduce their emissions of greenhouse gases, most especially carbon dioxide (CO_2) from fossil fuel combustion. In specific terms, Annex B countries pledged to reduce their

GHG emissions by 5.2% on average below aggregate 1990 emission levels between 2008-2012 their commitment period (UNFCCC, 1997). The agreement will not enter into force until the double-triggers conditions are met: Firstly, the national parliaments of at least 55 Parties to the Convention must ratify the treaty. Secondly, at least 55% of the total 1990 $_{CO2}$ emissions must be accounted for by industrialized countries that are amongst the ratifying parties.

Several controversial issues on the implementation of the Convention were outlined at the Kyoto Conference. The issues include credits for carbon sinks, i.e. agricultural soils and forests that store CO_2, and the question of restricted versus full tradability of emission rights across Annex B countries. In March 2001, the United States under President George W. Bush unequivocally refused to ratify the Protocol due to the huge cost it would have on the U.S. economy and due to the unacceptability of the exemption of developing countries such as China and India from binding emission targets.

The major difference between the Kyoto Protocol and the UNFCCC is that, while the UNFCCC admonished industrialized countries to steady GHG emissions, the Protocol obligates them to do so (because of its binding power). The recognition of the fact that developed countries are principal contributors to the current high levels of GHG emissions in the atmosphere as the result of over 150 years of industrial actions, the Protocol places a heavier emission reduction burden on them under the principle of "common but differentiated responsibilities." To date, 189 Parties of the Convention have ratified the Protocol.

The Kyoto Protocol has been hailed by a section of the intellectual community as a milestone in multilateralism. Proponents of the Protocol celebrate it as a landmark in international climate policy and celebrate it as a momentous achievement towards mitigating global warming. However, another point of view sees the approach in terms of its efficacy in the implementation of the agreement, namely setting timetables and targets for emission reductions as seriously defective (King, 2015).

Without any hesitation, the Protocol was a product of over 10 years of negotiations on climate policy. It became the first legally binding universal treaty on climate protection, and entered into force in 2005. The advocates of multilateralism celebrate the protocol as a breakthrough in international climate policy, because:

1. It envisaged considerable emission reductions for the developed world vis-à-vis emissions from their business activities.
2. It established an extensive global instrument for deepening and widening climate protection actions in the future.
3. It constituted the pioneer international environmental agreement that was built on market based mechanisms to determine cost-efficient reactions to the undoubted need for GHG reduction.

4. It designed a burden-sharing scheme set up for the first commitment period which ended in 2012 (Bohringer, 2003).

In the light of the above, proponents stated that despite an effective emission reductions in the initial commitment period, the ratification of Kyoto is important for the further policy process of climate protection. To them, Kyoto Protocol has been able to establish a flexible broad-based universal instrument that offers an appreciated starting point for modeling effective climate policies in the future (Bohringer, 2003).

On another hand, antagonists of the Kyoto Protocol avowed that it was bound to fail due to the flaws in its architecture. They concluded "that the Kyoto Protocol is an impractical policy focused on achieving an unrealistic and inappropriate goal" (McKibbin & Wilcoxen, 2002 p.127). Some of the key points advanced against the Kyoto Protocol goes as follows:

1. "The Kyoto Protocol is defective on both efficiency criteria (spatial and temporal equalization of abatement costs) because it omits a substantial fraction of emissions; (Thus failing the spatial criterion) and has no plans beyond the first period (thus not attending to the temporal dimension)" (Nordhaus, 2001 p.8);
2. "Kyoto does not deter free-riding and non-compliance" (Barrett, 1998 p. 38);
3. "The most fundamental defect of the Kyoto Protocol is that the policy lacks any connection to ultimate economic or environmental policy objective" (Nordhaus, 2001 p.13);
4. "The Kyoto Protocol has an arbitrary allocation of transfers. Moreover, since developing countries are omitted, they are completely overlooked in the transfers" (Nordhaus, 2001 p.9);
5. "The Protocol permit trading [as the principal policy instrument of the Kyoto Protocol] runs the risk of being highly inefficient, given uncertainties in the marginal cost of abating greenhouse gas emissions. This would probably generate large transfers of wealth between countries" (McKibbin & Wilcoxen, 2002 p.26);
6. Opponents to the Protocol have condemned it as a "deeply flawed agreement that manages to be both economically inefficient and politically impractical" (McKibbin & Wilcoxen, 2002 p.107); and
7. "No individual government has an incentive to police the agreement. The Kyoto Protocol can only work if it includes an elaborate and expensive international mechanism for monitoring and enforcement" (McKibbin & Wilcoxen, 2002 p.126).

In fact, years after the demanding negotiations for its implementation, the Protocol has not yet been implemented, even after Russia's ratification made of the number required for it to take effective legal status. The refusal of the United States to ratify the Protocol as well as the full tradability of emission rights granted to the former Eastern Bloc in surplus of its estimated future emissions imply that the Kyoto Protocol is very likely to achieve very little in terms of global emission reductions (Springer, 2002). This evolution seems to affirm the position of the antagonists of the Kyoto Protocol that its central approach of setting timetables and targets for emission reductions is seriously faulty (Bohringer, 2003).

COP 15 Copenhagen Climate Change Conferences

From 7 to 18 December 2009, the fifteenth Conference of the Parties (COP) of UNFCCC took place in Copenhagen. 120 Heads of State and Governments and over 50,000 participants participated making the Conference the utmost profile meeting of any multilateral environmental issue (Sindico, 2010). Before COP 15 international climate change debates have been along two parallel tracks. One was under the Kyoto Protocol and the UNFCCC. The first track was launched in Bali, Indonesia, at COP 13 of the UNFCCC, and countries were to devise ways to achieve "full, effective and sustained implementation of the Convention through long-term cooperative action, up to and beyond 2012." Under the second track, countries have been negotiating ways to enhance further Kyoto Protocol Annex I Parties' obligations (Sindico, 2010). Both tracks were to have ended at COP 15, but it was evident in the meetings preceding to COP15 and COP/MOP5 that this would not be achieved.

The Copenhagen Accord does not specify targets for GHG emissions reductions for any sector. It states that deep international emissions cuts are needed to hold the increase in global temperature to under two degrees Celsius. The Accord relies on industrialized nations to set their own economy wide emission reduction targets to take effect in 2020 by January 31, 2010. (Copenhagen Accord, 2009)

The Copenhagen Accord refers to very loose emission reduction essential to prevent a 2.0 degree rise in overall temperatures taking cognizance of the general goal of the UNFCCC. However, any specific cap was not establish. While COP 15 provided for a much more flexible approach as far as it gives each State, both developing and industrialized, the chance to decide its level of climate change mitigation level or goal, the Kyoto Protocol established a general cap baseline and also indicated what level of reductions each Annex I Party had to achieve between 2008 and 2012 commitment period. Thus, the environmental integrity of a regime based on the Copenhagen Accord would be contingent on whether the compliance with the emissions reductions level provided for would actually reduce an overall rise in temperatures to 2.0 degrees or even better, 1.5 degrees (Sindico, 2011).

The flexibility of the emission ceiling rate became a major weakness of COP 15 because countries of the EU in particular, were aiming at establishing new legally binding international treaty. This is because, a legally binding instrument will have enforceable obligations that are binding upon State Parties as well as a compliance mechanism that address situations of non-compliance. So it can be induced that COP 15 had no enforceable mechanism and lacked a strong compliance system.

Hopes were frustrated and amplified when, after two weeks of negotiations, a fairly small but influential group of countries led by the United States was able to negotiate the Copenhagen Accord (Sindico, 2010).

The compliance mechanism of the Copenhagen Accord has the ability to take actions against countries not complying with their commitments under the Kyoto Protocol while the nonexistence of a compliance structure in COP 15 offers states freedom for compromises. In the Copenhagen Accord "compliance" becomes "measurement, report and verification (MRV)" which seems to be designed in three diverse ways depending on whose mitigation action is considered. First, climate change mitigation action enshrined in pledges from Annex I countries will be "measured, reported and verified (MRV) in accordance with existing guidelines adopted by the Conference of the Parties". Second, mitigation action assumed by developing countries will be subjected to nationally established MRV. The Copenhagen Accord necessitates mitigation actions to be taken by Non-Annex I Parties vis-à-vis their domestic MRV. Bi-annually, the outcome will be reported through their national communications channels. Lastly, non-Annex I Parties can also choose to implement mitigation acts, which will be supported by international support. In this case, the Copenhagen Accord establishes that the national mitigation actions will be subjected to international MRV in agreement with rules adopted by the COP (Sindico, 2010).

The COP 15 can be said to be a failure in that it did not meet the expectation of the need of a binding mechanism that could replace the Kyoto Protocol or at best supplement it. The Copenhagen agreements were merely taken note of but not adopted.

The Accord also failed to address the major challenges of the Kyoto Protocol which is the failure to ensure that the US, the main emitter of greenhouse gas emissions (the US) ratifies it and the fact that emission reduction obligations was not set for evolving economies such as China, India and Brazil.

As such, the conference was said to be a failure for not achieving binding commitments to lessen global greenhouse gas emission levels adequate to meet the standards identified by about 3,000 prominent global scientists of the United Nations Intergovernmental Panel on Climate Change (IPCC) as measures to prevent calamitous costs such as food disruption, sea level rise leading to massive population displacement, tropical disease migration, water shortages, and obliteration of biodiversity (Allison & Bindoff et al., 2009).

From an environmental point of view, the Copenhagen Accord was a failure because, even if Parties complied with their pledges, the general rise in average temperatures will not be reduced to 2 degrees. Additionally, the flexibility Copenhagen Accord provided for countries in their climate change mitigation action was too much.

From a legal view point the Copenhagen Accord also fails because the obligations it provided for were voluntary while the MRV system shifted away from the compliance system existing in the Kyoto Protocol.

Finally, in spite of all the uncertainties that one may have on the legal, environmental and political flaws of the Accord, the undeniable fact is that it was the first time in about a decade that emerging developing countries and the US agreed to a framework to mitigate and adapt climate change in the future. The Copenhagen Accord was negotiated principally between the United States and the BASIC countries (Brazil, South Africa, India and China). Put differently, the involvement of key emerging countries of China, India, Brazil and South Africa and the United States, in the negotiation of the final Accord as well as the agreement by Mexico to host the next climate conference were quite significant. This is as a result of the prior decline of the countries in making commitments to greenhouse gas emission reduction for the Kyoto Protocol (Kampert, 2001).

While the convergence of 193 nations at Copenhagen to address the global climate problems were truly unique (Broder, 2009; Fahrenthold, 2009), the Copenhagen Climate Conference and its Accord have been generally regarded as a failure (Darren, 2009; Eilperin & Faiola, 2009; Garman, 2009).

COP 16, Cancun, Climate Change Conferences

The decision of the United States led by George Walker Bush to renege on the Kyoto agreement gave environmentalist hope that COP 16 United Nations Climate Change Conference in Cancun, Mexico, which took place from 29 November to 11 December 2010 would provide the platform to make the US commit to climate change.

Expectations for Cancun were modest, with few anticipating a legally-binding outcome or agreement on each outstanding issue. The key challenge that was faced by the countries of the world was to continue the process of constructing a sound foundation for meaningful, long term global action which was accomplished in Cancun.

Nevertheless, many still hoped that Cancun would produce meaningful progress on some of the key issues. In the lead-up to the conference, several matters were widely identified as areas where a balanced "package" of outcomes could be agreed. These issues included mitigation, adaptation, financing, technology, reducing emissions from deforestation and forest degradation in developing countries. The conservation, sustainable management of forests and enhancement of forest and its monitoring, reporting and verification (MRV) were also included. Negotiations on these key issues took place throughout the two-week meeting (IISD, 2010).

One of the major decisions resulting from COP 16 is to reduce deforestation and forest degradation through the creation of financial incentives to fund the conservation, sustainable use and enhancement of forest carbon stocks (REDD+). This decision on a REDD+ framework could have positive and/or negative impacts depending its application. There is broad consensus that REDD+ initiatives have the potential to conserve biodiversity and ecosystem services, improve local livelihoods, promote adaptation, and provide incentives to reform forest governance if well designed.

There is also broad recognition of the negative impacts REDD+ implementation could bring. Deleterious consequences might include the infringement of indigenous rights; introduction of invasive tree species (i.e. eucalyptus) to 'grow' CO_2 credits; ongoing degradation of natural forests leading to loss of biodiversity, species extinction and ongoing CO_2 emissions; labour and human rights abuses; destruction of plants relied upon by local communities for medicine and nutrition; loss of customary access to forests; resulting decline in nutrition and human health of forest dependent communities; and the disruption of ecosystems and loss of ecosystem services (Perron-Welch, 2011).

The reality of REDD+ lies in the particularities of each project and whether that project adequately balances environmental, social and economic factors to achieve a sustainable solution supported and enforced by law or voluntary certification. In this vein, it is important to closely monitor the financial underpinnings of this incentive scheme to ensure equitable and ecological outcomes rather than those based on speculation and fraud and lead to ongoing forest death. Thus, good governance is one key to successful implementation at all levels (Perron-Welch, 2011). REDD+ will necessarily be impacted by, and have an impact on, the way that international rules and commitments play out. To achieve the goal of reducing emissions from deforestation and forest degradation, the international community will need to understand the interplay between pre-existing rules and commitments on forests and those made at the Cancun conference (Perron-Welch, 2011).

The delegates in Cancun succeeded in writing and adopting an agreement that assembles pledges of greenhouse gas (GHG) cuts by all of the world's major economies, launches a fund to help the most vulnerable countries, and avoids some political landmines that could have blown up the talks, namely decisions on the (highly uncertain) future of the Kyoto Protocol. At the end only hope was dashed (Duruji & Duruji-Moses, 2016).

Assessing the Key Elements of the Cancun Agreements

In a nut shell, the 32-page Cancun Agreements made provisions for the following issues:

First, the Cancun Agreements provide for emission mitigation targets and actions (submitted under the Copenhagen Accord) for approximately 80 countries-including, importantly, all of the major economies.

The Agreements codify pledges by the world's largest emitters-including China, the United States, the European Union, India, and Brazil-to various targets and actions to reduce emissions by 2020. The distinction between Annex I and non- Annex I countries is blurred even more in the Cancun Agreements than it was in the Copenhagen Accord which does not signifies a step in the right direction. Also, for the first time, countries agreed – under an official UN agreement-to keep temperature increases below a global average of 2 degrees Celsius. This brings these aspirations, as well as the emission pledges of individual countries, into the formal UN process for the first time, essentially by adopting the Copenhagen Accord one year after it was "noted" at COP-15.

Another important aspect of the Accord hinges on the mechanisms for monitoring and verification that were laid out for "international consultation and analysis (ICA)" of developing country mitigation actions. Countries will report their GHG inventories to an independent panel of experts, which will monitor and verify reports of emissions cuts and actions.

Third, the Agreements establish a so-called Green Climate Fund to deliver financing for mitigation and adaptation. The World Bank was named as the interim trustee of the Fund despite the numerous objections from many developing countries, and even created an oversight board whose membership had about half of its members from the donor countries (Stavins, 2010)

More so, the developed countries on the platform of the Agreements agreed to mobilize $100 billion annually by 2020 to support mitigation and adaptation in developing countries. This fund would target sources from public and private resources (that is, carbon markets and private finance), bilateral and multilateral flows, as well as the Green Climate Fund. However, whether the resources ever grow to the size laid out in Copenhagen and Cancun will depend upon the individual actions of the wealthy nations of the world.

Fourthly, the Agreements advanced initiatives on tropical forest protection or better put, in UN parlance, Reduced Deforestation and Forest Degradation, or REDD+, by taking the next steps toward establishing a program in which the wealthy countries can help prevent deforestation in poor countries by possibly working through market mechanisms. This came despite exhortations from Bolivia and other leftist and left-leaning countries to keep the reach of "global capitalism" out of the policy mix.

Fifth, the Cancun Agreements established a structure to assess the needs and policies for the transfer to developing countries of technologies for clean energy and adaptation to climate *Change,* and a Climate Technology Center and Network (though yet undefined) to construct a global network to match technology suppliers with technology needs.

Also along this line was the fact that the Agreements endorsed an ongoing role for the Clean Development Mechanism (CDM) and other "market based mechanisms;" indicating that carbon capture and storage (CCS) projects should be eligible for carbon credits in the CDM an offer of special recognition for the situations of in Central and Eastern European countries (previously known in UN parlance as "parties undergoing transition to a market economy") and Turkey, all of which are Annex I countries under the Kyoto Protocol, but decidedly poorer than the other members of that group of industrialized nations.

It embraces parallel processes of multilateral discussions on climate change policies. Also, there was movement forward with specific, narrow agreements, such as on: REDD+ (Reduced Deforestation and Forest Degradation, plus enhancement of forest carbon stocks); finance; and technology. Such movement forward has, in fact, occurred in all three domains in the Cancun Agreements.

In the light of the above discussions, the parties to the Cancun meetings could maintain sensible expectations and thereby develop effective plans. It was able to create a long-term action strategy to address the threat of global climate change. The conference was therefore able to map out for the countries of the world a pathway for an effective plan of action for climatic change challenges humanity currently faces. According to Stavins, (2010) the successes of the Cancun conference enumerated above could be attributed to the following reasons particularly in contrast with the outcome of the Copenhagen conference.

First, the Mexican government through careful and methodical planning over the past year prepared itself well, and displayed tremendous skill in presiding over the talks. Mexican Minister of Foreign Affairs, Patricia Espinosa, the President of COP-16, carefully took note of the objections of Bolivia (and, at times, several other leftist and left-leaning Latin American countries, known collectively as the ALBA states), and then simply ruled that the support of 193 other countries meant that "consensus" had been achieved and the Cancun Agreements had been adopted by the Conference. She therefore explained that "consensus does not mean unanimity". The diplomatic role and acumen played by the president was widely applauded unlike the chairing of COP-15 in Copenhagen by Danish Prime Minister Lars Lokke Rasmussen, who allowed the objections by a similar same small set of five relatively unimportant countries (Bolivia, Cuba, Nicaragua, Sudan, and Venezuela) to derail those talks, which hence "noted," but did not adopt the Copenhagen Accord in December, 2009.

The key role played by the Mexican leadership is consistent with the notion of Mexico as one of a small number of "bridging states," which can play particularly important roles in this process because of their credibility in the two worlds that engage in divisive debates in the United Nations: the developed world and the developing world.

Second, China and the United States set the tone for many other countries by dealing with each other with civility. The tone of negotiations and discussion at COP 16 was braced with civility and respect unlike at Copenhagen where finger pointing by China and the United States dominated deliberation. As Elliot Diringer of the Pew Center on Global Climate Change wrote;

They may have recognized that the best way to avoid blame was to avoid failure. Beyond this, although the credit must go to both countries, the change from last year in the conduct of the Chinese delegation was striking. It appeared, as Coral Davenport wrote in The National Journal, that the Chinese were on a "charm offensive." Working in Cancun on behalf of the Harvard Project on Climate Agreements, I can personally vouch for the tremendous increase from previous years in the openness of members of the official Chinese delegation, as well as the many Chinese members of civil society who attended the Cancun meetings. (Stavins, 2010 p.3)

Third, a worry hovered over the Cancun meetings that an outcome perceived to be failure would lead to the demise of the UN process itself. This was because many nations (especially the developing countries, which made up the vast majority of the 194 countries present in Cancun) very much want the United Nations and the UNFCCC to remain the core of international negotiations on climate change, that implicit threat provided a strong incentive for many countries to make sure that the Cancun talks did not "fail."

Fourth, the negotiators continued a process for the construction of a sound foundation for meaningful, long term global action. The acceptance of the Cancun Agreements suggests that the international diplomatic community could be said to have recognized that incremental steps in the right direction are better than acrimonious debates over unachievable targets (like as it was in the Copenhagen conference).

According to Narain (2010) the Cancun conference was concluded in a deal in that it endorsed an arrangement that emission reduction commitments of industrialized countries will be decided on the voluntary pledge they make. They will tell us how much they can cut and by when. The US, which has been instrumental in getting the deal at Cancun, is the biggest winner because she is the largest emitter of greenhouse gases and will be free to cut emission at a convenient rate without having much negative impact on her economy if there will be.

The principle of equity in burden sharing prevailed at the conference. Under the Cancun deal, all countries including India and China are now committed to reduce emissions (a plus to COP 15) because of the operation of the principle. The pledge of the parties to the conference was to reduce energy intensity by 20-25 per cent by 2020 which is part of the global deal. At the start of the meeting nobody would have envisaged that the burden of the transition or change could shift to the developing world but the conclusion was of all countries bearing willingly a percentage of GHG control that they are capable of curbing within the ceiling fixed.

COP 17 Durban 2010 Climate Conference

The past two decades have witnessed attempts at global agreements on tackling climate change which have been futile and this includes the problems at the Copenhagen conference and the tussle to save the multilateral climate regime in Cancun (Ares, 2012). Many decisions were postponed until Durban which included the decision on what will succeed the Kyoto protocol. The United Nations Climate Change Conference in Durban, South Africa took place in 2011,

At Durban, the Parties decided "to launch a process to develop a protocol, another legal instrument or an agreed outcome with legal force under the Convention applicable to all Parties." The Durban platform had a mandate of negotiating a new climate agreement which will be "applicable to all." COP 17 was resolute on concluding a new agreement by 2015 and enters into force in 2020. Negotiations for a new agreement was characterised by much disagreements over the roles of the developing countries and the industrialised ones. The dispute was on the removal of the distinction and obligation between Annex I (developed countries) and non-Annex I countries (traditionally developing countries) that was instituted in the Framework Convention of 1992. The argument of Annex 1 countries is that the obligation imposed on the industrialised countries by the Convention and the Kyoto Protocol have not existed for less industrialised countries and the distinction was outdated. They argued that the reasons for the distinction was no longer valid because the developing countries are now wealthier than the traditional industrialised countries and there contribution to the global emission has also increased rapidly (Obergassel, Lukas, Mersmann, Ott, and Wuppertal, 2016: 8). The new formula therefore for another legal instrument was a compromise between, the EU and many developing countries on the one hand and on the other hand the US, China and India who insisted that there should be no new commitments for developing countries (Obergassel, Lukas, Mersmann, Ott, and Wuppertal, 2016).

Despite the disagreements, a number of feats were achieved at Durban. One of the achievements of the Durban summit is the decision reached to extend the Kyoto Protocol after the end of its first commitment period in 2012. This allowed for the continuity of the existing tools and mechanisms which are the Clean Development Mechanism, the registries and Joint Implementation until 2017 or 2020.

The Durban platform for Enhanced Action was also established. The Protocol will be prepared by the Ad Hoc Working Group on the Durban Platform for Enhanced Action (ADP). The new Protocol would be legally binding and applicable to all Parties (Erbach, 2015). The Platform also included a second period of commitment of the Kyoto Protocol which kicked off on January 2013 and would also avoid the gaps in the first commitment period. Russia, USA and Japan however refused to sign up to the second Kyoto Protocol commitment period. New rule on forestry were approved to advance the Protocol's environmental veracity (Euroclima, 2012).

The initiative of the EU and the Alliance of Small Island States (AOSIS) was able to agree on identifying options for closing the "ambition gap" between the objective of ensuring global warming is kept below 2°C and current emissions reduction pledges for 2020. The establishment of new market based mechanisms was another giant stride achieved at COP 17. The new market mechanisms were established to reduce emission and complement the Clean Development Mechanism (CDM) as well as enhance the cost effectiveness of actions. Climate change issues related to agriculture were also discusses with the view of arriving at a policy decision at the end of 2012 (Euroclima, 2012).

COP 17 established a Technology Mechanism to facilitate technology development, transfer adaptation and finance (Erbach, 2015). The conference at Durban established the Technology Executive Committee that had the responsibility of providing analysis, recommendation and linking various institutions related on developing technologies as well as its transfer. The Climate Technology Centre and Network was charged with matching the technological needs of developing countries. An Adaptation Committee will provide parties technical supports; disseminate information; analyse information made available by the Parties and also provide recommendations on adaptation (Oberthur, Antonio, Vina and Morgan, 2015)

A new platform for financing a green climate was established. The new Green Climate Fund was set up as a major medium for promoting low-carbon climate resilient growth through multilateral climate financing (Euroclima, 2012). Other acheivements include the formation of a new working group to formulate a new climate design or framework involving all countries by 2015 for implementation by 2020. For the first time, emerging economies especially China was willing to discuss emission reduction goals to be implemented in 2020 (Morel, Bellassen, Deheza, Delbosc and Leguet, 2011).

The Durban Package marks a constellation of international, national and sub-national institutions and actors with expertise capacity and authority to address climate change. The Durban conference therefore marked a breakthrough in international efforts to combat climate change.

COP 21 Paris 2015 Climate Conferences and Paris Accord Ratification

After two decades of global climate efforts, UN Framework Convention on Climate Change, the COP 21 Paris 2015 provided a platform to arrive at a landmark agreement to chart a new course in the global climate change comparable to Kyoto agreement. The outcome in Paris was a four years negotiation rounds which commenced at the 17th Conference of the Parties in Durban 2011. The Agreement eventually ratified on 12 December 2015 by 196 parties to the UN Framework Convention on Climate Change (UNFCCC) was a milestone in international climate policy and multilateral deliberations. The Paris Accord ratification is a multilateral treaty in which both developing and developed countries agreed to take responsibility which will be entrenched in their national context yet geared towards a global goal of holding temperature "well below" 2°C while also pursuing effort to stay below 1.5°C (Climate focus, 2015).

The agreement also represents a landmark in distinguishing between developed and developing countries. It was slightly able to move beyond the 1992 Kyoto agreement which established sternly distinct commitments for the two groups of countries. Paris agreement supplemented the principle of common but differentiated responsibilities with its fundamental obligations for all parties and by its array of procedures used to institute difference between parties (Bodle, Donat and Duwe, 2016).

The Paris Agreement defines a universal, legal framework to 'strengthen the global response to the threat of climate change' (Art. 2) (C2ES, 2015). The essential legal obligations are procedural. The Agreement does not specify mitigation actions or the timeline for the achievement of emission levels. Rather it focuses on individual national climate mitigation plans that would be achieved via a transparency framework. That is, for the first time, on the history of climate change diplomacy, each country will develop its own plan for climate change mitigation and then communicate the plan and contribution to the secretariat of the Convention. It give Parties a five-year 'cycles', to prepare voluntary 'nationally determined contributions' (NDCs), report on implementation of NDCs, justify their contributions and frequently improve the strategy in the light of a universal stocktake. In the light of this, developing countries will continue to get support to pursue their NDCs and in reporting their climate change situations. Consequently, one accomplishment of the Paris Accord is its heavy reliance on the cooperation and determination of national efforts pursued on transparency and regular reports of progress and achievements (Bodle, Donat and Duwe, 2016). It also provided a platform for international review of both the strategies and actual achievements of the NDCs.

The agreement also pays cognisance to the different responsibilities and starting points of countries. Article 2.2 accentuates that the Agreement will be implemented in harmony with the 'principle of common but differentiated responsibilities and respective capabilities' which applies 'in the light of different national circumstances.' This implies that the fight against climate change will be led by the developed countries, which also will in turn; provide support to the actions taken by the developing countries in this direction.

In spite of the challenges in presenting its legal details, clearly presents political narrative on its objectives and the actions expected of Parties to achieve them. It provides direction on the flow of finance and technology. Like COP 17, developed countries shall provide financial and capacity building support and information on technology transfer to developing countries who will make information available on the support received and more needs (Arts. 9-11).

The Paris Accord also explored the strengths of interdependence, transparency and sincerity amongst nations to drive global emission reduction. Article 13 provides an 'enhanced transparency framework for action and support' that will offer a distinct insight of mitigation action. The Paris Agreement was designed to thrive on openness and exchange of information amongst the Parties

However, the setback for the Accord just like Kyoto happened in America where a party that is skeptical of climate change was again elected to replace the one that galvanized the world to reach the agreement

The administration of Donald Trump argues that US compliance with the Paris accord could "cost America as much as 2.7 million lost jobs by 2025, according to the National Economic Research Associates." Why risk that when the Paris Agreement would lead to only a minuscule reduction in global temperature. Trump further argued that in "14 days of carbon emissions from China alone would wipe out the gains from America's expected reductions in the year 2030." Another reason advanced by Trump is that America remaining in the agreement would cost the US economy "close to $3 trillion in lost GDP and 6.5 million industrial jobs, while households would have $7,000 less income, and in many cases, much worse than that." To Donald Trump the US participation would require the US to pay a significant sum about $3 billion to the Green Climate Fund that was set up by the accord when many of the other countries have not spent anything. And many of them will never pay one dime.

Though most of the reasons cited by the United States president has been rebuffed, the action of the US present a danger to the future implementation of the Paris agreement by parties (Variensky, 2017)

IS MULTILATERALISM DEAD?

A system can be defined a set of components with identifiable attributes, among which patterned relationships persist over a period of time. It is composed of parts or units that are independent, acting individually but interrelated as they all function for the central purpose of the success or the welfare of the system as a whole. The international system is the environment in which international actors (typically states) interact. It is made up of parts that are hierarchical as some are more significant than others, interacting with each other to produce an outcome. In the international political system, these parts are nation-states, international organizations, and several other entities that have power of an international scale.

The enduring features of the international political system are perpetuation of the territorial nation-state, the corresponding support for the principle of sovereignty, reliance on self-help measures (rather than a supranational authority) to achieve national political measures. Derivation of international legal norms and obligations from both custom and formal consent, acceptance of the pursuit of power through preparation for war, structural inequalities through the persistence of various hierarchies - economic, political and resource hierarchies, as well as military asymmetries independence, cooperation, dependence, sanction and underdevelopment are the hallmark of international system (Kegley & Wittkopf, 1999).

Thus the characteristic of the international system defines how actors would act in the face of challenges of common effects like terrorism, climate change and it attendant consequences. Put differently, the features of the international system like interdependence, dependence, cooperation, and sanction have become permanent characteristics of the structure of relations in world politics. States will continue to be confronted by calamities beyond their capacities or on a large scale.

Multilateralism is not dead but is in coma. The pursuit of the national interest of a state is the driving force of a country's foreign policy. Therefore states worry about a division of possible gains that may favour others more than itself. Though the Westphalia state is waning, the worry by state parties of becoming dependent on others and losing their autonomy still reinforce resistance to multiterism.

The world faces old and new security challenges that are more complex for a single state to handle. This has placed a demand on the principle of cooperation and interdependence on the various actors in the international system. International cooperation is ever more necessary in meeting these challenges however; many are unilaterally acting to cut emissions in ways that are consistent with the development priorities of their individual states (Smith, 2009). In his Nobel lecture, President Obama acknowledged the importance of multilateralism when he stated that the world must come together to confront climate change: "There is little scientific dispute that if we do nothing, we will face more drought, more famine, more mass displacement – all of which will fuel more conflict for decades" (Obama, 2009).

In spite of the general belief that the world need to act in a concerted manner to tackle climate change some of the issues articulated below (some are mentioned above) still militate against the collective combating of climate change. First are the uncertainties on the cost and benefit of GHG abatement that render decision-making in climate policy very difficult. Second, there are only weak economic and political mechanisms to enforce cooperative behavior between sovereign countries (Bohringer, 2003). The third point is the lack of a supranational authority that could coerce countries into the implementation of globally efficient climate policies. The main challenge to climate policy is thus to shape international agreements that create incentives for sovereign states to enter cooperation (Bohringer, 2003).

Countries that benefit less from cooperation would definitely not comply because it is an investment in an unproductive or unprofitable business (Botteon & Carraro, 1993).

The fourth major challenge of achieving multilateral objective on climate change is that countries especially the developing ones fear goal of the collective bodies. There remains a fear that the global environmental goals are being set according to the agenda of countries in the global industrialized north, particularly the United States that on two different occasions constituted a wedge to solving global problems using multilateral platform.

CONCLUSION

Fundamentally, the creation of international institutions by states is to create a means of achieving collective objectives that cannot be achieved individually. As such, the preservation of the already disappearing global environment demands some devolution of sovereign powers under a recognized multilateral institution.

Though countless negotiations after the breakthrough in Kyoto left many disillusioned on the capacity of multilateral institution to build a global consensus and facilitate nation-state commitment on implementation, the triumph of environmental activists and state parties in Paris 2015 and the subsequent ratification of the agreement is still dependent on the vagaries of political outcome in the US. However, the action of the Trump administration to pull US out of the agreement, can only be a temporary setback. This is apt given the unison of the rest of the world to forge ahead with the Paris accord even without the United States.

REFERENCES

Adams, A. (1993). Food Insecurity in Mali: Exploring the Role of the Moral Economy. *IDA Bulletin*, *24*(4), 41–51. doi:10.1111/j.1759-5436.1993.mp24004005.x

Ares, E. (2012). *Durban Climate Conference*. Science and Environment Section SN/SC/6140.

Armstrong, D., Lloyd, L., & Redmond, J. (2004). *International Organization in World Politics. Hound mills*. Basingstoke, UK: Palgrave Macmillan. doi:10.1007/978-0-230-62952-3

Barrett, S. (1998). Political Economy of the Kyoto Protocol. *Oxford Review of Economic Policy*, *14*(4), 20–39. doi:10.1093/oxrep/14.4.20

Bindoff, Bindschadler, Cox, de Noblet, England, Francis, … Weaver. (2009). The Copenhagen Diagnosis: updating the world on the latest climate science. The University of New South Wales, Climate Change Research Centre (CCRC).

Bodle, R., Donat, L., & Duwe, M. (2016). *The Paris Agreement: Analysis, Assessment and Outlook. Background paper for the workshop "Beyond COP21: what does Paris mean for future climate policy?" 28 January 2016, Federal Ministry for the Environment, Nature Conservation, Building and Nuclear Safety*. Berlin: BMUB.

Bohringer, C. (2003). *The Kyoto protocol: A Review and Perspective*. ZEW discussion paper No. 03-61, Mannheim.

Broder, J. M. (2009a, December 19). Many Goals Remain Unmet in Five Nations Climate Deal. New York Times, pp. 2-3.

Broder, J. M. (2009b, December 17). Poor and Emerging States Stall Climate Negotiations. New York Times, pp.1-2.

Bump, P. (2017). Nine reasons Trump's withdrawal from the Paris climate agreement doesn't make sense Politics Analysis. *Washington Post*. Retrieved from https://www.washingtonpost.com/news/politics/wp/2017/06/01/all-the-reasons-that-trumps-withdrawal-from-the-paris-climate-agreement-doesnt-make-sense/?utm_term=.8128f4219a0d

Caroline, B., & Peterson, J. (2009). *Conceptualizing Multilateralism*. Mercury Working Paper.

Climate focus. (2015). *The Paris Agreement: Summary*. Climate Focus Client Brief on the Paris Agreement III 28 December 2015. Briefing Note. Retrieved from http://unfccc.int/paris_agreement/items/9444.php

Cooper, A. F. (2002). Like-minded nations, NGOs, and the changing pattern of diplomacy with in the UN system: An introductory perspective. In A. F. Cooper, J. English, & R. Thakur (Eds.), *Enhancing Global Governance: Towards a New Diplomacy?* Tokyo: United Nations University Press.

Darren, S. (2009, December 21). Obama Negotiates 'Copenhagen Accord' With Senate Climate Fight in Mind. *The New York Times*, p. 1.

Duruji, M., & Urenma, D.-M. (2016). The Environmentalism and Politics of Climate Change: A Study of the Process of Global Convergence through UNFCCC Conferences. In *Handbook of Research on Global Indicators of Economic and Political Convergence*. Hershey, PA: IGI Global.

Elliot, J. A. (1999). *Sustainable Development*. London: Routledge.

Erbach, G. (2015). *Negotiating a new UN climate agreement: Challenges on the road to Paris. EU*. European Parliamentary Research Service.

Euroclima. (2012). *Key outcomes of the Durban climate change*. Retrieved from http://www.euroclima.org/en/euroclima/our-people/item/655-principales-resultados-de-la-cumbre-sobre-cambio-clim%C3%A1tico-de-durban

Fahrenthold, D. A. (2009, December 19). Copenhagen Climate Talks, by the Numbers. *Washington Post*.

Falola, A. (2009, December 19). Climate Deal Falls Short of Key Goals. *Washington Post*, p. 4.

Garman, J. (2009, December 20). Copenhagen-Historic Failure That Will Live in Infamy. The Independence, pp. 3-4.

Hampson, F. O., & Reid, H. (2003). Coalition Diversity and Normative Legitimacy in Human Security Negotiations. *International Negotiation*, 8(1), 7–42. doi:10.1163/138234003769590659

IISD. (1997). UN Climate Change Conference. *Earth Negotiation Bulletin*.

Kagan, R. (2002). Power and Weakness. *Policy Review, 113*. Retrieved 22nd September, 2011 from the website www.policyreview.org

Kampert, P. (2001, April 17). U.S. Takes Heat; why is Bush's Stand on Global warming Treaty Upsetting Nations around the World. *Chicago Tribune,* pp. 2-4.

Keck, M. E., & Sikkink, K. (1998). *Activists beyond Borders*. Ithaca, NY: Cornell University Press.

Kegley, C. W., & Wittkopf, E. R. (1999). *World Politics: Trend and Transformation*. New York: World Publishes.

Keohane, R. (2006). The Contingent Legitimacy of Multilateralism. *Garnet Working Paper, 9*(6).

Keohane, R. O. (1986). Reciprocity in International Relations. International Organization, 27, 1-27.

Keohane, R. O. (1990). Multilateralism: An Agenda for Research. *International Journal (Toronto, Ont.), 45*(4), 731–764. doi:10.1177/002070209004500401

Keohane, R. O., & Nye, J. S. (2000a). Introduction. In Governance in a Globalizing World. Washington, DC: Brookings Institution.

Keohane, R. O., & Nye, J. S. (2000b). Power and Interdependence. New York: Addison-Wesley Longman.

King, E. (2015, February 2). Kyoto Protocol:10 Years of the Worlds First Climate Change treaty. *Climate Home News*. Retrieved from http://www.climatechangenews.com/2015/02/16/kyoto-protocol-10--years-of-the-world-first-climate-change-treaty/

Kunnas, J. (2009). The Theory of Justice in A Warming Climate. *Earth Environ. Science,* (6), 11. doi:10.1088/1755-1307/6/1/112029

Martin, L. (1992). Interests, Power and Multilateralism. *International Organization, 46*(4), 765–792. doi:10.1017/S0020818300033245

Mather, A. S., & Chapman, K. (1995). Environmental Resources. London: Longman.

McKibbin, W. J., & Wilcoxen, P. J. (2002). The Role of Economics in Climate Change Policy. *The Journal of Economic Perspectives, 16*(2), 107–129. doi:10.1257/0895330027283

Narain, S. (2010). *The Cancun End Game, Bad For Us Bad For Climate. IIED Outreach, A multi stake holder magazine on environment and sustainable development*. Retrieved from http://www.field.org.uk/files/outreach_cs_cancun_outcomes.pdf

Nordhaus, W. D. (2001). *After Kyoto: Alternative Mechanisms to Control Global Warming*. Academic Press.

Obama, B. (2009a). *Nobel Lecture*. Retrieved from http://nobelprize.org/nobel_prizes/peace/laureates/2009/obamalecture_en.html

Obama, B. (2009b). *Remarks by the President at United Nations Secretary General Ban Ki-Moons Climate Change Summit.* Retrieved from http://www.un.org/wcm/webdav/site/climatechange/shared/Docume

Obergassel, W. (2016). Phoenix from the Ashes—An Analysis of the Paris Agreement to the United Nations Framework Convention on Climate Change. Academic Press.

Oberthür, S., Viña, A., & Morgan, J. (2015). *Getting Specific On The 2015 Climate Change Agreement: Suggestions For The Legal Text With An Explanatory Memorandum.* Working Paper.

Perron-Welch, F. (2011). *The Future of Global Forests after the Cancun Climate Change Conference IDLO Sustainable development law on Climate Change.* Legal working paper series.

Presentation at the 20th Anniversary Meeting of the International Energy Workshop, Romano, GC. (2010). *The EU-China Partnership on Climate Change: Bilateralism Begetting Multilateralism in Promoting a Climate Change Regime?* Mercury, paper No 8. Pp 1-28.

Ruggie, J. G. (1992). Multilateralism: The Anatomy of an Institution. *International Organization, 3*(46), 561–598. doi:10.1017/S0020818300027831

Shanaha, M. (2010). *Journalist from Climate Change Frontline Kept World's Eyes Focused on COP 16. IIED Outreach, A multi stake holder magazine on environment and sustainable development.* Retrieved from http://www.field.org.uk/files/outreach_cs_cancun_outcomes.pdf

Sindico. (2010). The Copenhagen Accord and the Future of the International Climate Change Regime. *Revista Catalana de Dret Ambiental, 1*(1), 1 – 24.

Singer, P. (2002). *One World: The Ethics of Globalization.* New Haven, CT: Yale University Press.

Smith, A. (2009). *The Nobel Peace Prize 2009: Time for Hope.* Retrieved from http://nobelprize.org/nobel_prizes/peace/laureates/2009/speedread.html

Springer, U. (2002). The market for tradeable GHG permits under the Kyoto Protocol: A survey of model studies. *Energy Economics, 25*(5), 527–551. doi:10.1016/S0140-9883(02)00103-2

Stavins, R. N. (2010). *A Look Back At Cop-16. What happened (and why) in Cancun. IIED Outreach, A multi stake holder magazine on environment and sustainable development.* Retrieved on September, 28, 2011 from the website http://www.field.org.uk/files/outreach_cs_cancun_outcomes.pdf

Turner, R. K. (1988). *Sustainable environmental management.* London: Belhaven, UNFCCC 1992. U.N. Doc. A/AC.237/18, reprinted in 31 I.L.M. 849.

UNFCCC. (2009). *Copenhagen Accord.* Retrieved from http://www.denmark.dk/NR/rdonlyres/C41B62AB-4688-4ACE-BB7BF6D2C8AAEC20/0/copenhagen_accord.pdf

Varinsky, D. (2017). 5 claims Trump used to justify pulling the US out of the Paris Agreement—and the reality, Tech. *Business Insider.* Retrieved from http://www.pulse.com.gh/bi/tech/tech-5-claims-trump-used-to-justify-pulling-the-us-out-of-the-paris-agreement-and-the-reality-id6773413.html

WCED (World Commission on Environment and Development). (1987). Our Common Future. Oxford University Press.

Werksman, J. (1995). Greening Bretton Woods. In The Earthscan Reader in Sustainable Development. London: Earthscan.

Yasmine, R. (2010). COP 15 and Pacific Island States: A Collective Voice on Climate Change. *Pacific Journalism Review*, *16*(1), 193–203.

Section 2
Regional Politics

Chapter 3
Water Security in Pakistan

Sofia Idris
Independent Researcher, Pakistan

ABSTRACT

Pakistan largely faces water scarcity, and the arid agricultural land of Pakistan is mainly due to the violation of the Indus Water Treaty by its neighbor India. By blocking the river flow towards Pakistan from the head-works, India has been building excessive dams, barrages, and power projects which are illegal according to the above-mentioned international treaty, and Pakistan has many times appealed in the UN to take action against this unfair act. However, so far, nothing could be done in this regard. The study will be helpful to understand the various challenges facing Pakistan to cater the insufficient supply of water and will give insight on the most important dimensions and facts about the international challenges to meet the shortage. The trans-boundary water issue between China and India have also been studied to try to explore new options and find the solution of a much pressing problem. The study might thus contribute to understand the issue, study the role of international community, and give useful and practical suggestions to solve the most pressing problem.

INTRODUCTION

The floods in Pakistan this year was nothing new and nobody would have probably been surprised. However, the destruction that it brought with it was surely disappointing; people would have surely been disappointed for flooding happens every year, but no preventive measures have been taken by any government in 67 years. Pakistan largely faces water scarcity where there is shortage of it even in the urban areas that includes large and important cities too. The poor quality of water also adds to the problem where hundreds and thousands of children die each year because of the water related diseases. Underground water reservoirs are being used up presently to cater the needs of its people, which is dangerous for the excessive extraction of water

DOI: 10.4018/978-1-5225-3990-2.ch003

from underground water beds while not being recharged and may result in water logging and salinity in the soil resulting in barrenness and ultimately desertification of the land of a country which has agriculture based economy.

The arid agricultural land of Pakistan is mainly due to the violation of the Indus Water Treaty by its neighbor, India. By blocking the river flow towards Pakistan from the head-works, India has been building excessive dams, barrages and power projects which are illegal according to the above mentioned international treaty and Pakistan have many times appealed in the UN to take action against this unfair act. However, so far, nothing could be done in this regard.

Pakistan has objected on the construction of Baglihar HPP; which is the run-of-the-river power project on Chenab and was conceived in 1992. Under the Treaty, India cannot reduce the flow in Chenab River below 55,000 cusecs between June 21 and August 31 where Pakistan received as low as 20,000 cusecs during August/September 2008. Pakistan objected on the construction of the Baglihar dam over the River Chenab and feared about the expected blockage and a resultant shortage in its water share which seems to be coming true. Swiss engineer appointed by the World Bank, suggested a few design changes; this proposal was to help limit flow control. The two countries agreed they would honour the verdict, but India has failed to make the necessary design alterations. The conflict is still unresolved with the issue of Kishanganga HEP which is a run-of-the-river hydroelectric scheme designed to divert water from the Kishanganga river __ Neelum river in Pakistan to a power plant in the Jhelum river basin. Its construction began in 2007 and has been completed in 2016 (The News, 2017). World Bank being the broker and the signatory of the treaty has not been successful yet in terms of resolving the conflict.

But the construction of dams in India is not the only problem for Pakistan; the unexpected and sudden release of river water flow from India at the times of heavy monsoon rains adds to problem when the medium level or low-level floods become heavy and dangerous floods adversely affecting Pakistani citizens as well as its economy. Many experts have stressed on building the dams and barrages as a best solution to overcome these problems; hence, it is high time to focus on the construction of projects like the long forgotten Kalabagh dam and look for ways to settle the problem for a sustainable future.

For instance, experts participating in "International Conference on Water Resources Governance in the Indus Basin" arranged by the Department of Political Science, GC University & United Nations University, Institute for Water, Environment and Health in 2013 were of the view that Himalayas is undergoing extensive construction of dams. Moreover, it is expected that Afghanistan would construct a lot of dams on Kabul River just like India for the latter has been convincing it to do so. India has developed a lot due to construction of dams and in contrast; Pakistan has failed to do so. Thus, Pakistan need to build more and more dams especially Kalabagh Dam.

Therefore, the people are worried how to meet their everyday water needs while the annual monsoon rains spoil everything: spoil the crops and floods ruin the property and take many lives while displacing many families. The floods also adversely affect the country's economy every year. The government has the responsibility to take all the necessary steps to protect its people from such catastrophes and prevent such happenings from taking place in the future. Many scholars and experts call for construction of dams and barrages in this regard. This would not only cater the annual water needs of Pakistan particularly during the times when there is no rain and water is scarce in the country, irrigate the crops and produce electricity while combating energy crisis.

Meanwhile the recent tension on the Line of Control (LoC) at the Indo-Pak border followed by terrorist attacks in the Pakistan's province of Baluchistan and the massive killings, torture and coercion by India in the disputed Indian Occupied Kashmir (IOK) have changed the course of international politics where India, once to be dubbed the most favorite nation in terms of trade with Pakistan is now hostile with its neighbor. Therefore, Pakistan is now in a state of war. The situation has worsened to the level that India has now violated the 56-year-old Indus Water Treaty that it had signed with Pakistan in 1960 and threats to continue violating it and even isolate Pakistan diplomatically. India has too accused Pakistan of sponsoring terrorist attacks in India but has so far failed to provide its evidences to the international community. The two nuclear states are now heading towards another war which is considered to be more destructive than ever.

The study will be helpful to understand the various challenges facing Pakistan to cater the insufficient supply of water and will give insight on the most important dimensions and facts about the international challenges to meet the shortage. It will also study a case of trans-boundary water issue between China and India at the Brahmaputra River in order to explore new options and find a solution of a much pressing problem. The study might thus contribute to understand the issue, study the role of international community and give useful and practical suggestions to solve the most pressing problem.

BACKGROUND

Pakistan receives low and variable rainfall and is an arid to semi-arid economy. Agriculture occupies the largest part of the economy as majority of its people are directly or indirectly associated with it. Water resources are scarce in the country. The agriculture sector uses most of the water resources of the country. Indus Basin irrigation system is a major water resource, which irrigates more than 90 per cent of the country crops. Therefore, water is quite important for economic growth and employment (Gollnow, 2016).

The Partition of the Indian subcontinent in 1947 cut the Indus River system into two troublesome irrigation developments. An understanding on water sharing between Pakistan and India was clearly necessary. Therefore, the Indus Waters Treaty (IWT) was brokered by the World Bank in 1960 which split six shared rivers three a-piece between Pakistan and India instead of water flows in individual tributaries, the detailed discussion of which have been provided later in the chapter (see Table 1) (Label, Xu, Bastakoti & Lamba, 2010; Wirsing, 2008; Zawahri, 2009).

The geographical characteristics of a boundary or a territory, where these make up vital resources, play a role in inducing conflict. Boundaries are lines of possibility for both cooperation and conflict since they can impact deeply on human physical, economic and social well-being. If the demarcation of borders does not facilitate the achievement of these goals for the states on both sides of the boundary, the borders themselves can become a cause of conflict. As according to Waterman, the root cause of the hardships of the peoples of Palestine and Ireland lies in superimposed boundaries which inadequately take into account geographical realities (Waterman, 1984). In the same way, the boundary drawn in Punjab provided an opportunity to India to exploit cross-boundary water resources as military weapon and well as a motivation to capture Kashmir territory. This brought about the Indus water dispute which ended in an international war between Pakistan and India in 1948. Additionally, the armistice boundary in Kashmir, determined under the UNSC ceasefire of 1948/1949, created a permanent situation of suspense which has filled an enormous sense of insecurity in equally the Pakistani state and public, who see themselves vulnerable since the Jammu and Kashmir territory remains under Indian control (Mirza, 2016).

Table 1. Salient features of Indus Waters Treaty (IWT)

PAKISTAN	INDIA
UNRESTRICTED USE OF RIVERS GRANTED	
Chenab	Beas
Jhelum	Sutlej
Indus	Ravi
RESTRICTIONS AND PERMISSIONS	
International financial assistance to Pakistan for utilizing the waters of western rivers	Limited permission to build storages on the western rivers
	Restriction of extension of irrigation development in India

Michael Klare's observation provides additional support to this fact, who views the Indian obduracy in retaining control over the Jammu and Kashmir region, linked with water-politics of India not to give up upper riparian status and was overshadowed by political, ideological, political and military dimensions. Klare contends that Indian upper riparian status in Kashmir have enormous political implications for the future course of regional politics as well as the future use of the Indus rivers (Klare, 2002). It also establishes the Kashmir dispute as a conflict of realistic interests, based on the vital Indus water resource, on the one hand, as well as the status of Kashmir as a hydro-strategic territory on the other. The findings of Lipschutz further strengthen the argument that scarcity is a product of resource control and not of the particular attributes of nature (Mirza, 2016; Lipschutz, 1997).

Pakistan's political climate has probably become most aggressive to India's endeavors to carry out more projects in the upper reaches of the basin's eastern rivers while such suspicions also exist in both Nepal and Bangladesh toward India's objectives in building these projects. Additionally, Pakistan, as a country that is mostly dependent upon a single river basin, the Indus, and its tributaries, it is in precarious position for numerous reasons. Firstly, its economy is more or less entirely reliant upon water-hungry sectors, including the wheat, sugar, and textile industries, and secondly, its poorly maintained and aging water infrastructure leaves the country apt to worsening power shortages. Both these problems have been caused by an extensive canal system whose control of the headwaters is with India. In these circumstances, the development of hydropower in Kashmir, on both the Pakistan and Indian side, has become an ever more significant issue. In fact, some have contended that water resources have in fact been an important element of the disputes historically, albeit to some extent it has been ignored in the public discourse on the region (Hill, 2012).

IMPACT OF WATER SCARCITY

Water scarcity, inefficient water management, population explosion, and dependency on agricultural production have made the country particularly vulnerable to climate change.

Groundwater is being extensively used to meet water necessities of crops due to limited surface water availability. Consequently, several environmental issues arise such as a falling water table and an increase in secondary salinisation of irrigated soils. Moreover, the usability of groundwater is limited in many areas, since it is already saline or polluted because of insufficient waste and wastewater treatment (Gollnow, 2016).

It is estimated that climate change effects, including heat stress, water scarcity, and rainfall shifts, may globally prompt migration anywhere from 25 million to 1 billion people by 2050. Climate change together with other socioeconomic factors is also linked with the rural to urban migration trend (Gollnow, 2016).

INDUS WATER TREATY (IWT)

Pakistan is one of the world's most arid countries and is deeply dependent on a yearly influx into the Indus River system. Mostly, water in the system emanates from the bordering country and is typically derived from snow-melt in the Himalayas. Therefore, country's hydraulic economy faced massive challenges since its inception in 1947.

The water dispute surfaced in April 1948, after India closed the canals on the eastern rivers, Ravi and Sutlej, and agreed to reopen them only after the provisional agreement, the Inter Dominion Agreement of May 1948, claiming the entire water of eastern rivers. In 1960, Indus Water Treaty (IWT) was signed by the two countries under the mediation of World Bank. The treaty was a remarkable achievement as it brought to an end the long-standing water dispute between Pakistan and India. IWT's primary objective was to fix and delimit the rights and obligations of the water usage of each country in relation to other. According to IWT, Pakistan got the western rivers (Chenab, Jhelum, and Indus) and India, the eastern rivers (Beas, Sutlej, and Ravi). Moreover, India was not permitted to build storages on the western rivers but to a very limited extent. Other restrictions included extension of irrigation development in India. IWT also had provisions regarding the extent of irrigated agriculture, the exchange of data on project operation, etc. IWT also provided certain institutional arrangements; these include, a permanent Indus Commission which would consist of a commissioner each for Pakistan and for India, and there were to be exchanges of visits and periodical meetings. The treaty also gave provisions for the resolution of any differences that might arise. Furthermore, IWT also included the stipulation to provide international financial assistance to Pakistan for utilizing the waters of western rivers for the development of irrigation works (Iqbal, 2010).

The Indus Water Treaty (IWT) has XII articles with Annexure A - H. The Article I of the treaty is concerned with the definitions while the Article II is about the provisions regarding the Eastern Rivers. The Article IV focuses on the provisions regarding the Western Rivers. Article V is about the financial provisions. Article VI states about the exchange of data. The Article VII gives provisions related to future cooperation. Article VIII is about the provisions regarding the creation of the post of the Permanent Indus Commission. Article IX is about the settlement of differences and disputes. Article X gives emergency provision. Article XI focuses on general provisions. Article XII states the final provisions. Next, the IWT includes

Annexures from A to H. Annexure A is about the exchange of notes between Government of India and Government of Pakistan. Annexure B is concerned with the agricultural use by Pakistan from certain tributaries of the Ravi. Annexure C is about the agricultural use by India from the Western Rivers. Annexure D focuses on the generation of Hydro-electric power by India on the Western Rivers. Annexure E is about the storage of waters by India on the Western Rivers. Annexure F is about the Neutral Expert. Annexure G is about Court of Arbitration and finally, Annexure H focuses of transitional arrangements (World Bank, n.d.).

Pakistan and India agreed on the Indus Waters Treaty in 1960 that was brokered by the World Bank. According to IWT, the three Western Rivers of the subcontinent; Indus, Jhelum and Chenab were given unrestricted availability to Pakistan whereas India was given the unrestricted availability of the three Eastern Rivers; Ravi, Sutlej and Bias.

PRESENT SITUATION

Recently released report on Pakistan's water security by United Nations Development Programme (UNDP) indicates that the water situation in the country is quite alarming and the efforts to address it insufficient. The report shows that Pakistan ranks better only than Ethiopia with only 121 cubic meters of per capita planned live water storage capacity per person (Pakistan Today, 2017).

The report further elaborates that the present government has failed to ensure water security satisfactorily since avoidable delays in the Neelum-Jhelum Hydroelectric Power Project has enabled India to complete its Kishanganga Dam upstream so as to legally secure a larger portion of the water on Neelum which means that Neelum-Jehlum will not be getting the required amount of water downstream to generate the expected 969 megawatts.

Some of the factors for the worsening water security are increasing population and users of water multiplying and depleting reserves of ground water. But, the report considers the lack of attention to water scarcity by successive governments including military dictatorships as the real issue (Pakistan Today, 2017).

Moreover, the report indicates that a failure to increase water availability by no less than 14.2% will result in Pakistan not being capable of meeting its demand by 2025. With increasing effects of global warming every year it is frightful to note that water availability has dropped by almost 806 cubic meters per inhabitant since 1990 (Pakistan Today, 2017).

The water profile of the country has changed drastically for the reason that it developed from being a water-abundant country to a water-stressed country. During the period 1990–2015, per capita water availability fell from 2,172 to 1,306 cubic metres per citizen. Pakistan extracts nearly 75 per cent of its "freshwater annually, thereby exerting tremendous pressure on renewable water resources. Despite remarkable improvements in the proportion of the population using improved water sources and improved sanitation facilities", over 27 million people still do not have access to safe water and about 53m do not have access to sufficient sanitation facilities. Around 39,000 children under the age of five die each year from diarrhea caused due to unsafe water and poor sanitation (Artaza, 2017).

TRANSBOUNDARY WATER ISSUES

As mentioned earlier, there have been a lot of conflicts or disagreements on water with its neighbor, India. Most of these conflicts or disagreements between India and Pakistan are about the building of big dams. The building of dams has consequences for the environment, the entire region and its people. Both the countries do not have the permission to obstruct the flow of water as according to the Indus Waters Treaty. Also the Indus Waters Treaty (IWT) states that they should take each other into account whilst planning and eventually building on one of the rivers. One of the main provisions in the treaty was the exchange of information, but in fact, this does not seem to be totally enforced. One of the most significant conflicts is about the Baghlihar dam. In the conflict, India is supposed to be the offender of the treaty. (Keetelaar, 2007)

There have been several other conflicts and discussions related to the provisions in the treaty. For instance, the Salal Hydroelectric Project, in which Pakistan objected to the sluice gates. Agreement between both governments on this project was reached after India removed the construction of the sluice gates from the project. Also a debate existed about the height of the dam of the Dul Hasti Project when Pakistan raised objections (Sharma, 2006). The work on the Kishanganga dam and the Wullur BarrageTulbul Navigation Project was called to a halt by Pakistan. Pakistan claims they would obstruct the water flow to Pakistan as these projects violate the Indus Waters Treaty (IWT). Since 1988, talks on both of these projects have been taking place via the composite dialogue but they still disagree (Deccan Herald, 2005; Alam, 2013; The Nation, 2016). Moreover, the conflicts that already exist between Pakistan and India are on the border demarcation in the provinces of Jammu and Kashmir; these conflicts over water put an additional tension on the relationship among both countries in general. Nonetheless India has a lot of other planned hydro power projects that could bring about more discussion under the conditions of the treaty (Husain 2005).

BAGLIHAR HYDROPOWER PROJECT

In 1992, The Baglihar Hydropower Project was conceived by India, and its construction began by 1999. The work was planned on the Chenab River, one of the rivers that were allocated to Pakistan through the IWT. According to this treaty, India is obliged to report to Pakistan about the accurate details of any such plan sited on any of the Pakistani rivers. In addition, thereupon wait for approval by Pakistan since it has only restricted permission to develop such a project. (Keetelaar, 2007)

The Clause 1 & 2 of Article III of the Indus Waters Treaty (IWT) states that:

1. *Pakistan shall receive for unrestricted use all those waters of the Western Rivers which India is under obligation to let flow under the provisions under the paragraph (2).*
2. *India shall be under an obligation to let flow all the waters of the Western Rivers, and shall not permit any interference with these waters, except for the following uses, restricted (except as provided in item (e) (ii) of Paragraph 5 of Annexure C) in the case of each of the rivers, The Indus, The Jhelum and The Chenab, to the drainage basin thereof:*
   a. *Domestic Use;*
   b. *Non-Consumptive Use;*
   c. *Agricultural Use, as set out in Annexure C; and*
   d. *Generation of hydro-electric power, as set out in Annexure D." (World Bank, n.d.)*

Moreover, the Clause 4 of Article III of the Indus Waters Treaty (IWT) states that:

4. *Except as provided in Annexures D and E, India shall not store any water thereof, or construct any storage works on, the Western Rivers (World Bank, n.d.).*

The main argument by Pakistan against the project therefore, is that the Baglihar Hydropower structure entailed gated spillways hence creating the opportunity for storage and the creation of a reservoir. Moreover, Pakistan speculated that there is the possibility of India to use this gated structure in a war like situation to either hold back the water supply or flood Pakistan. On the contrary, India claimed that the project did not involve water storage, but that it was a run-of-the river project, and would provide benefits for the local people. Permanent Indus Commission could not succeed in its attempt that had been made to negotiate about the project (Hali, 2004).

Therefore in 2005, Pakistan approached the World Bank to assign a neutral expert in concordance with one of the arbitration provisions in the treaty. Raymond Lafitte, the Swiss water expert was appointed. He was initially expected to give his decision in November 2006 but this was nonetheless postponed until February 2007.

India desired to have a bilateral dialogue to resolve the disagreement. Subsequently, Pakistan demanded to halt the work on the project until a verdict has been reached. Nevertheless, Lafitte did not agree with this, since this would cost too much money. It was the first time in 45 years of history to have executed the arbitration clause of the Indus Waters Treaty (Rediff, 2005).

Under the Treaty, India cannot reduce the flow in Chenab River below 55,000 cusecs between June 21 and August 31 where Pakistan received as low as 20,000 cusecs during August/September 2008. Pakistan objected on the construction of the Baglihar dam over the River Chenab and feared about the expected blockage and a resultant shortage in its water share which seems to be coming true. Swiss engineer appointed by the World Bank, suggested a few design changes; this proposal was to help limit flow control. The two countries agreed they would honour the verdict, but India has failed to make the necessary design alterations. The differences on the initial filling of the Baglihar dam (in Doda district in Jammu and Kashmir) have however been resolved in a spirit of cooperation and goodwill in 2008 (Parsai, 2010).

KISHANGANGA HEP

Kishanganga HEP is a run-of-the-river hydroelectric scheme designed to divert water from the Kishanganga river __ Neelum River in Pakistan to a power plant in the Jhelum river basin. Its construction began in 2007 and has been completed in 2016 (The News, 2017).

A 24-kilometre-long tunnel is to be used to divert the waters of the Kishanganga River for power production. The surplus water flow will join the Wullar Lake and eventually evade the 213 KM long Neelum River while it runs through Jhelum to Muzaffarabad.

The Kishanganga HEP is another case of serious violation of Article – IV (3), C of the IWT and especially paragraph (5) and Article VII (b). The stipulation of these articles clearly prevents India to increase the catchment area of any artificial or natural drain and drainage, beyond the area on the effective date. The diversion will increase the catchment area therefore resulting into violation of the treaty under the above-mentioned articles. As a result of diverting the flow of Kishanganga River, upstream at Gurez, it will increase the catchment area of river Jhelum (tributary of main Jhelum).

This may cause enormous material damage to occur due to adverse effects of non-availability/reduction of water in the Neelum valley. Secondly, this project will increase likelihood of floods resulting in erosion of agriculture land along both sides of River Jhelum tributary and material damage caused because the catchment area of river Jhelum tributary will increase the flow in tributary. In October 2011, the construction was halted by The Hague's Permanent Court of Arbitration owing to the protest by Pakistan against the resultant multifarious effects of the construction of the dam such as water shortage by 61% approximately, reduction in energy generation in winter season by 35%, economic loss of 141.3 USD million dollars annually and loss of already planned agricultural development activities in the area worth 421 million rupees (Naz, 2014).

A resolution was unanimously adopted by a rare joint sitting of the two committees of the National Assembly of Pakistan seeking immediate suspension of work on two disputed projects by India, Kishanganga and Ratle hydro projects and constitution of an arbitration court to resolve the ongoing water dispute. It is the responsibility of the World Bank to play its role without further delay according to IWT. Indian move to construct 45 to 60 dams on the western rivers has impelled Pakistan to have serious reservations (Khan, 2017).

Pakistan and World Bank will have a second round of talks on July 31, 2017 in Washington. Demand to appoint Chairman, Member Technical and Member Legal of the mediation/arbitration court on the conflict on Kishanganga and Ratle projects will be made on this occasion (Jang, 2017).

PERMANENT INDUS COMMISSION (PIWC)

The Permanent Indus Commission (PIWC) that provides a contemporary mechanism for conflict resolution and consultation using, inspection, exchange of data, and visits between both the countries has registered a list of 155 hydropower projects dams, in violation of the Indus Waters Treaty of 1960, India plans to construct on Indus, Jhelum and Chenab, an alleged move to deprive Pakistan of its water rights (Bhutta, 2011).

A list registered by PICW discovered that India had built 41 hydropower projects in addition to 12 hydropower plants were under construction, apart from the 155 projects drafted on the Western Rivers (Bhutta, 2011).

India is over and done with the construction of with 690-MW Salal 2, 6 hydropower plants on River Chenab, and 450-MW Baglihar 1. Construction on two projects was in progress, consisting of the 15MW Ranja-Ala-Dunadi and 450-MW Baglihar 2.

Besides, India has schemed an extra 56 hydropower projects on River Chenab, as well as some big projects for example the 1200-MW Sawalkot (1 and 2), 1020-MW Bursar (1 and 2), 1000-MW Pakaldul (1 and 2), 690-MW Rattle (1 and 2), 715-MW Seli, and 600-MW Kiru.

India has finished 15 projects on River Jhelum, in addition to the 105-MW Lower Jhelum and 480-MW Uri-1, 105-MW Upper Sindh (Bhutta, 2011).

Nevertheless, the International Court of Arbitration (ICA) has disqualified India from any permanent constructions on the controversial Kishanganga project responding on Pakistan's appeal for 'interim measures' with reference to the dam which may block the restoration of the river flow to its natural channel. Numerous projects put up by India have been objected to by Pakistan, in addition to the Baglihar dam on River Chenab. However, the decision was given against Pakistan when the issue was raised before a neutral expert against its erection (Bhutta, 2011).

One more dam violating the 1960 Indus Waters Treaty (IWT) had been designed on the Chenab River in the region of the Held Kashmir which is named as 1,380 MW Kirthai hydropower project. Kirthai hydropower project is intensifying environmental issues in Pakistan and intimidating energy rights and water security (Mustafa, 2014).

The situation continues to get worse instead of getting any better with the current tension on the Line of Control (LoC) at the Indo-Pak border followed by terrorist attacks in the Pakistan's province of Baluchistan and the massive killings, torture and coercion by India in the disputed Indian Occupied Kashmir (IOK) have changed the course of international politics where India, once to be dubbed the most favorite nation in terms of trade with Pakistan is now hostile with its neighbor. Therefore, Pakistan is now in a state of war. The situation has worsened to the level that India has now violated the 56-year-old Indus Water Treaty that it had signed with Pakistan in 1960 and threats to continue violating it and even isolate Pakistan diplomatically. India has too accused Pakistan of sponsoring terrorist attacks in India but has so far failed to provide its evidences to the international community. (The Times of India, 2016; The Express Tribune, 2017; Farooq, 2017)

In 2016, China blocked a tributary of the Brahmaputra River in Tibet, which sent agitation to India. The action was part of the building of China's most expensive and the long-planned hydro Lalho project in Xigaze. The move was in time when Indian PM Narendra Modi's threatened to scrap the Indus Water Treaty. Although China had assured that the construction would not interrupt water flow toward India; nevertheless, many considered the move as a "soft" message that India should abstain from instigating water wrangles with Pakistan. (Fazil, 2017; The Express Tribune, 2016; Khadka, 2016)

CHINA: ANOTHER COUNTRY WITH TRANSBOUNDARY WATER ISSUES

China is the most important upstream country for ecological security and transboundary water in Asia with sharing 110 international lakes and rivers along its northeast, southwest, northwest, and southwest borders as well as being home to a good number of Asia's great rivers that flow into eighteen downstream countries. This geographic position makes the trans-boundary water resources along with corresponding environmental issues important elements of China's regional and international relations. China's entire trans-boundary water resources (TWR) of about 800 billion m3 in 2006 make up 31·72% of total runoff in China (Ministry of Water Resources of the People's Republic of China's 2007), a large amount of which start off in the southwest of China, predominantly from the 'Asian Water Towers' on the Tibetan-Qinghai Plateau. These waters have not been heavily used and are still high quality – both rarities given China's economic growth in addition to intensive use of natural resources. More prominently, these waters have an effect on the lives of over two billion people in China in addition to 10 downstream countries (Immerzeel, van Beek & Bierkens, 2010; He, D., Wu, R., Feng, Y., Li, Y., Ding, C., Wang, W. and Yu, D. W., 2014).

China and India share, although not exclusively between them, four major rivers. The Indus/Shiquan River is shared by Pakistan, China and India. The Brahmaputra River is shared by Bangladesh, Bhutan, India, and China. The Ghaghara and Kosi rivers are shared by Nepal, China, and India. Clearly, China and India do not share any rivers exclusively: all transboundary rivers of China and India are shared with other neighboring countries too (Zhang, 2016).

BRAHMAPUTRA WATER CONFLICT

One of Asia's major rivers is the Brahmaputra. The headwater of the Yaluzangbu-Brahmaputra River originates in Tibet Autonomous Region of China where it is known as the Yarlung Tsangpo and flows east before cutting through the Yaluzangbu Grand Canyon, then it flows down to India before it enters Bangladesh where it joins the Ganges and finally flows into the Bay of Bengal. The river is called "Jamuna" in Bangladesh, "Brahmaputra" in India, and "Yaluzangbu" in China (Khadka, 2017; Mahapatra & Ratha, 2016; Jamwal, 2013; Zhang, 2016; Huang, 2015).

The river data issue between China and India rose after the two countries ended a tense more than two months long stand-off over a disputed Himalayan border area.

Every year, the Brahmaputra gets severely flooded during monsoon season, causing huge losses in Bangladesh and northeast India. Both the countries have agreements with China that oblige the upstream country to share hydrological data of the Brahmaputra during monsoon season every year between 15 May and 15 October. The data is primarily about the water level of the river in order to alert downstream countries in the case of floods (Khadka, 2017; Mahapatra & Ratha, 2016; Jamwal, 2013; Zhang, 2016).

The conflict over water between the two Asian giants is the vital task for sustainable management of the basin. The conflict is alarming over inter-basin water transfers and water diversions, act of interfering with natural river flows from dams, sustained growth in water and energy demand. The basin is characterized by great seasonal fluctuation in water availability as a result of the extremely dry winter and the very wet monsoon.

The curious amalgam of fast economic growth, intensified global competition for energy resources and growing populations, is driving both India and China to put emphasis on hydropower. The planned water diversion and hydropower projects, along with mounting water security interests have a greater effect on relations between these two countries (Mahapatra & Ratha, 2016).

In addition, both India and China are rapidly growing economies, and are contending for the access to the same but limited water resources in the Asian sub-continent. China's need of usable water resources is already leading to a considerable shortfall in the annual GDP, with the potential to further worsen the situation with the persistent economic growth, and by means of the negative effects that are associated with the climate change within China. On the other hand, with an estimated population of 1.4 billion by 2050, India is also calculated to be "water-scarce" approximately during the same time (Jamwal, 2013).

Numerous scholars and analysts have speculated over the past decade about the likelihood of China and India going to war over water. Some assert that a future "water war" will occur—and others declare such fears overblown.

The glaciers in Tibet in China are melting at a faster rate, in addition to the growing water scarcity and coupled with a widening north-south regional water gap, China will be challenged with increasing pressure to implement a contentious upstream water diversion plan in its western provinces. This strategy will threaten India in view of the fact that the downstream part of the Brahmaputra River run through a disputed area with strong consequences for national sovereignty. Both countries will then increase their security footing in an already heavily militarized border region. As China's economic growth maintains its downward trajectory, popular nationalism will pressurize the Chinese Communist Party's capability to pursue a foreign policy uninfluenced by public opinion and populism. The possible net result: an expected water war between China and India (Pak, 2016).

MANAGING THE TRANSBOUNDARY WATER ISSUES

No Trans-Boundary Water Treaty

Unlike the Indus Water Treaty (IWT), there is currently no existing comprehensive bilateral or multilateral agreement governing the water management of the river, a water treaty regarding a fair use of water of the Brahmaputra River either on a sub-basin or all-basin level. Nonetheless, the two countries have founded Expert Level Mechanism (ELM) and in October 2013, India and China signed a Memorandum of Understanding in order to strengthen the cooperation on trans-border Rivers; however it is more of an agreement on China's specific commitment to sharing hydrological information of the Brahmaputra/ Yaluzangbu in monsoon season rather than a comprehensive bilateral mechanism (The News, 2016; The Express Tribune, 2016; Huang, 2015).

China agreed to extend the data provision period from May 15 to October 15 of the relevant year, in the new October Memorandum of Understanding, starting sixteen days earlier than the previous one. The memorandum acknowledges the importance of trans-border rivers cooperation in developing mutual strategic trust and communication. The Memorandum sets a good framework for future comprehensive cooperation even though it is not a comprehensive bilateral water treaty; it however, specifically addresses sharing some hydrological data. Thus, according to the Memorandum, the countries will provide hydrological data of the Brahmaputra/ Yaluzangbu River during flood seasons rather than the development activities. This will be almost equally effective for the reason that the hydrological information for the period of flood season will be a direct reflection of the operation of the dams. This may perhaps be a way to cooperate exclusive of having to disclose information which has been regarded as sensitive by the parties. In the memorandum, India has appreciated for China providing assistance in emergency management, extending the data-providing period as well as providing the hydrological data (Huang, 2015).

In addition, twenty-five water experts from Bangladesh, China, Nepal and India met in Dhaka In 2010 in order to discuss the future cooperation over the Himalayan River Basin and formulated The Dhaka Declaration. Chinese experts with little official background had participated alongside the former officials from India and Bangladesh. The Dhaka Declaration emphasizes the need to defend equal interests, especially those of the lower riparians, joint research projects, the serious consequences of climate change and the data sharing and scientific exchange (Huang, 2015).

Both India and China have engaged in strategies including the use of inclusive rhetoric, to prevent problems from escalating. Chinese leaders have emphasized the narrative of two great civilizations working all together to defend the interests of the Third World. Chinese rhetoric and actions concerning the Brahmaputra have tended towards "desecuritizing" water conflicts in an attempt to assuage Indian concerns. Chinese officials have been maintaining that China's dams on the upper reaches of the Brahmaputra as run-of-the-river which is not designed to hold water. India has received repeated reassurances from top Chinese leaders that Beijing has no intent of diverting the waters of the Brahmaputra River and that the Chinese would do nothing that would adversely affect India's interests (Ho, 2017; The News, 2016).

Over the last decade, China has made an effort to reduce two major Indian concerns in connection with the Brahmaputra: potential Chinese development activities along the river; and flooding that could be prevented with access to Chinese data. A lot of the concerns about flooding grew as a result of a major flood that occurred in June 2000. Many in India claimed that China withheld hydrological data that may possibly have prevented the disaster; this led to hostility in Sino-Indian relations. In the reaction to Indian concerns about flooding, India and China have established a series of agreements to interchange hydrological data. China agreed to share the hydrological data with India in April 2002 between June 1 and October 15 every year, corresponding to the yearly flood season. In November 2006, both India and China agreed to set up an expert-level group in order to discuss emergency response measures and hydrological data (Samaranayake, Nilanthi; Limaye, Satu & Wuthnow, Joel, 2016).

Secondly, China has sought to alleviate Indian concerns over Chinese development activities all along the river. China's public rhetoric has been largely unsuccessful to assuage Indian concerns. While Indian officials have not overtly rejected Chinese pledges that the Tibetan dam-building won't harm the Indian interests, India's official position has been similar to that adopted by the United States of America in its arms control negotiations with the Soviet Union in the 1980s: "Trust but verify" (Samaranayake, et al., Joel, 2016).

Another challenge for China is related to the Indian efforts to build on the Brahmaputra in Arunachal Pradesh. Arunachal Pradesh is one of the two major dispute areas along the Sino-Indian border. Indian infrastructure development along the Brahmaputra is of special concern for China for the reason that it could grant India leverage in border negotiations in addition to setting hurdles for Chinese efforts to gain control of this territory. Besides sovereignty concerns, Chinese observers also identify environmental risks caused by Indian development of the river. One Chinese claim, is that Indian industrial activity in Arunachal may well increase sedimentation of the river, which may possibly raise the danger of flooding in parts of Tibet. Other Chinese sources emphasize that rising Indian carbon emissions related to greater industrial activity in the region could be a factor to glacial melt in the Himalayas, and endanger the long-term flow of the river. One approach used by China in the recent years has been to leverage its influence in international institutions for instance, the Asian Development Bank to reject it's funding to India for infrastructure projects in the disputed area (Samaranayake, et al., 2016; Jianxue, 2010; Li, 2013; Garver, 2001; Zhifei, 2013; Ramachandran, 2009; Doshi, 2016).

In the mid-June of 2017, China and India were caught in a row. The Brahmaputra River water dispute started when a tense more-than-two-months-long stand-off over a disputed Himalayan border area ended. (Khadka, 2017; BBC News, 2017) The move was in time when Indian PM Narendra Modi's threatened to scrap the Indus Water Treaty (IWT). Although China had assured that the construction would not interrupt water flow toward India; nevertheless many considered the move as a "soft" message that India should abstain from instigating water wrangles with Pakistan (Fazil, 2017; The Express Tribune, 2016; Khadka, 2016).

RELATING PAKISTAN AND CHINA'S EXPERIENCE AND RECOMENDATIONS

Both Pakistan and China are challenged with trans-boundary water issues with India. China has these issues at the Brahmaputra River whereas Pakistan faces such problems at all the rivers since all of them originate from Kashmir (a disputed region between Pakistan and India) which then flows through India and then Pakistan; it then ultimately empties into the Arabian Sea.

There is no comprehensive bilateral or multilateral agreement governing the water management of the Brahmaputra River. Nevertheless, India and China have founded Expert Level Mechanism (ELM) and in October 2013, the two countries signed a Memorandum of Understanding in order to strengthen the cooperation on trans-border Rivers. It is more of an agreement on China's specific commitment to sharing hydrological information of the Brahmaputra/ Yaluzangbu in monsoon season rather than a comprehensive bilateral mechanism (The News, 2016; The Express Tribune, 2016; Huang, 2015).

But India and Pakistan signed Indus Water Treaty (IWT) in 1960 under the mediation of World Bank. The treaty was a remarkable achievement as it brought to an end the long standing water dispute between Pakistan and India. IWT's primary objective was to fix and delimit the rights and obligations of the water usage of each country in relation to other. According to IWT, Pakistan got the western rivers: Chenab, Jhelum and Indus and India, the eastern rivers: Beas, Sutlej and Ravi (Iqbal, 2010; Label, Xu, Bastakoti & Lamba, 2010; Wirsing, 2008; Zawahri, 2009).

There are many reasons of water related conflict between China and India as well as Pakistan and India. Firstly, India accused China of withholding the hydrological data which is the violation of their agreement. Furthermore, the water diversion/ obstruction of the flow of water are also part of the conflict between China and India.

On the other hand, India has itself been caught in illegal development in violation of the Indus Water Treaty IWT. The Permanent Indus Commission (PICW) discovered that India had built 41 hydropower projects in addition to 12 hydropower plants were under construction, apart from the 155 projects drafted on the Western Rivers (Bhutta, 2011). Pakistan also faces the problem of water diversion/ obstruction of the flow of water by India. Moreover, Indian development activities are intimidating energy rights and water security. Both India and China make developments at the Brahmaputra River which offends the other.

Both these trans-boundary water issues are for military & strategic reasons. Indian development activities are intensifying environmental issues in Pakistan. Hence, the resultant water scarcity and security issues that arise include several environmental issues such as a falling water table and an increase in secondary salinisation of irrigated soils (Gollnow, 2016). Development at the Brahmaputra River by India, is also increasing environmental risks including sedimentation of the river, flooding, carbon emissions leading to glacial melt in the Himalayas, and endanger the long-term flow of the river (Samaranayake, et al., 2016).

Nevertheless, the major element which determines the position of power in both these trans-boundary water security issues is being the upper riparian and China enjoys this position against India whereas, Pakistan being the lower riparian with India, face much more difficulties. Therefore, India is more compliant with China even though it also has strategic issues at one of the two major dispute areas along the Sino-Indian border, Arunachal Pradesh with Aksai Chin being the other (Samaranayake, et al., 2016; Garver, 2001). China being the upper riparian can be observed to have been constantly trying to assuage Indian concerns. However, India is intransigent when it comes to water-politics with Pakistan while it constantly violates IWT (Mirza, 2016; Lipschutz, 1997). Additionally, Pakistan, as a country that is mostly dependent upon a single river basin, the Indus, and its tributaries, it is in precarious position for numerous reasons (Hill, 2012).

With Table 2, it can be deduced that India has a comparatively more radical approach with Pakistan albeit the nature of the problem is almost the same. The international arbitration and mediation has either halted or altered the construction designs of India on the Pakistani waters, but it never permanently stopped India's endeavors to carry out more projects while it also managed to later complete the earlier unfinished projects, which added an insult to the injury. Hence, it is high time for the World Bank to make a permanent or at least a more effective settlement of the conflict. It should now consider drafting an effective mechanism, system or agreement so as to promote justice, peace, harmony, stability, and international cooperation to promote environmental sustainability and combat issues like climate change and environmental degradation.

Table 2. Comparison of trans-boundary water issues between China and Pakistan

CHINA	PAKISTAN
INTERNATIONAL TREATY	
No treaty or agreement; only Expert Level Mechanism (ELM) & Memorandum of Understanding signed in 2013	Indus Waters Treaty (IWT) signed in 1960
Agreement included exchange of information	Agreement included exchange of information
	Established a Permanent Indus Commission
POWER POSITION	
Upper riparian	Lower riparian
RIVALRY	
India	India
INDIAN ATTITUDE	
Defensive attitude being a lower riparian	Intransigence linked with water-politics of India not to give up upper riparian status.
REASONS FOR CONFLICT	
1. Withholding the hydrological data by China	Illegal development by India in violation of the treaty. PICW discovered that India had built 41 hydropower projects in addition to 12 hydropower plants were under construction, apart from the 155 projects drafted on the Western Rivers.
2. Development at the Brahmaputra River by both sides	Indian development activities are intimidating energy rights and water security
3. Military & strategic reasons	Military & strategic reasons
4. Development at the Brahmaputra River by India increasing environmental risks including sedimentation of the river, flooding, carbon emissions leading to glacial melt in the Himalayas, and endanger the long-term flow of the river	Indian development activities are intensifying environmental issues in Pakistan
5. Water diversion/ obstruct the flow of water	Water diversion/ obstruct the flow of water by India
6. Arunachal Pradesh is one of the two major dispute areas along the Sino-Indian border	Border demarcation in the provinces of Jammu and Kashmir

World Bank should take a stricter role in order to ensure a long-term peace and sustainability in the region. It may also put penalty and sanctions upon India, if necessary. Moreover, international financial assistance for such projects may also be denied to India just like Asian Development Bank (ADB) did in the favor of China. Furthermore, other countries may stress and isolate India diplomatically to pressurize it to set aside its unjustified obduracy against Pakistan just like when China blocked a tributary of the Brahmaputra River in Tibet, it created a difference for Pakistan in particular and the entire region in general. This is all necessary in order to promote global cooperation and environmental sustainability for a peaceful, prosperous and a sustainable future for it is a collective responsibility towards a sustainable environment leading to a sustainable future and those who are not conscious and/or do not have it listed in their priorities, will have to be convinced towards it. Whereas, many scholars and experts have stressed on building the dams and barrages as a best solution to overcome these problems; hence, it is high time to focus on the construction of projects like the long forgotten Kalabagh dam and look for ways to settle the problem for a sustainable future. This would not only cater the annual water needs of Pakistan particularly during the times when there is no rain and water is scarce in the country, irrigate the crops and produce electricity while combating energy crisis.

CONCLUSION

Pakistan is facing serious water scarcity and security issues. The country is experiencing these problems from both internal and external factors. The world today has developed from limited international cooperation to becoming a global village where not only business and technological assistance is readily available but also assistance and cooperation can be seen on complicated issues like terrorism and environmental sustainability. However, there are still some cases where such issues like environmental sustainability can become a matter of international competition rather than of international cooperation. The chapter is a study of such an example where Pakistan is faced with the challenge of international competition with India rather than of compliance and cooperation on the issue of water security which is causing water scarcity and environmental degradation like problems. A similar case study on China has also been included in an attempt to learn new options and find the solution to this latent problem. China has trans-boundary water issues with India on the Brahmaputra River.

On certain instances, India has violated the Indus Waters Treaty (IWT) signed by the two countries in 1960 which was facilitated by the World Bank. According to IWT, the three Western Rivers of the subcontinent; Indus, Jhelum and Chenab were given unrestricted availability to Pakistan whereas India was given the unrestricted availability of the three Eastern Rivers; Ravi, Sutlej and Bias. But over the years, there have been many instances of violation of the treaty on the part of India. The Permanent Indus Commission (PIWC) created according to the provisions of the IWT, discovered that India had built 41 hydropower projects in addition to 12 hydropower plants were under construction, apart from the 155 projects drafted on the Western Rivers, in violation of the Indus Waters Treaty of 1960, in an alleged move to deprive Pakistan of its water rights. Many scholars and experts have stressed on building the dams and barrages as a best solution to overcome these problems; hence, it is high time to focus on the construction of projects like the long forgotten Kalabagh dam and look for ways to settle the problem for a sustainable future. This would not only cater the annual water needs of Pakistan particularly during the times when there is no rain and water is scarce in the country, irrigate the crops and produce electricity while combating energy crisis.

The tension is still there and India continues to violate the treaty while the International Court of Arbitration (ICA) and the World Bank haven't been able to settle the serious and lingering conflict on water. Instead of things getting any better, the relationship between the two states have taken a hostile turn where there are desultory battles on and off at the border. Hence, it is high time for the World Bank to make a permanent or at least a more effective settlement of the conflict. It should now consider drafting an effective mechanism, system or agreement so as to promote justice, peace, harmony, stability, and international cooperation to promote environmental sustainability and combat issues like climate change and environmental degradation.

Both the trans-boundary water issues between China and India as well as that between India and Pakistan are for military & strategic reasons. Indian development activities are intensifying environmental issues in Pakistan. Hence, the resultant water scarcity and security issues that arise include several environmental issues such as a falling water table and an increase in secondary salinisation of irrigated soils. Development at the Brahmaputra River by India have also been increasing environmental risks including sedimentation of the river, flooding, carbon emissions leading to glacial melt in the Himalayas, and endanger the long-term flow of the river.

Nevertheless, the major element which determines the position of power in both these trans-boundary water security issues is being the upper riparian and China enjoys this position against India whereas, Pakistan being the lower riparian with India, face much more difficulties. Therefore, India is more compliant with China even though it also has strategic issues at one of the two major dispute areas along the Sino-Indian border, Arunachal Pradesh with Aksai Chin being the other. Therefore, it is suggested that World Bank should put penalty and sanctions upon India, if necessary. Moreover, international financial assistance for such projects may also be denied to India just like Asian Development Bank (ADB) did in the favor of China. Furthermore, other countries may stress and isolate India diplomatically to pressurize it to set aside its unjustified obduracy against Pakistan. This is all necessary in order to promote global cooperation and environmental sustainability for a peaceful, prosperous and a sustainable future for it is a collective responsibility towards a sustainable environment leading to a sustainable future and those who are not conscious and/or do not have it listed in their priorities, will have to be convinced towards it.

FURTHER STUDY

It is suggested that a more detailed study can be made on the works violating the Indus Waters Treaty (IWT). Furthermore, the current conflict on IWT can be studied comprehensively to get a better understanding of the issue; which will also give an insight on the water-politics on the other border that is, with Afghanistan. Moreover, a detailed study can be carried out along the lines of the international water conflict between Pakistan and India are deeply associated with the issue of Kashmir, a territorial conflict which started right after the inception of both the countries in 1947.

REFERENCES

Alam, U. (2013). *International Law and Freshwater: The Multiple Challenges.* Cheltenhem, UK: Edward Elgar Publishing.

Artaza, I. (2017, March 12). Water Security. *Dawn.* Retrieved from: https://www.dawn.com/news/1319986

BBC News. (2017, August 28). *China claims victory over India in Himalayan border row.* Retrieved from: http://www.bbc.com/news/world-asia-41070767

Bhutta, Z. (2011, November 15). Water wars: India planning 155 hydel projects on Pakistan's rivers. *The Express Tribune.* Retrieved from: http://tribune.com.pk/story/292021/water-wars-india-planning-155-hydel-projects-on-pakistans-rivers/

Doshi, V. (2016, May 18). India set to start massive project to divert Ganges and Brahmaputra rivers. *The Guardian.* Retrieved from: https://www.theguardian.com/global-development/2016/may/18/india-set-to-start-massive-project-to-divert-ganges-and-brahmaputra-rivers

Farooq, U. (2017, November 16). *Mumbai's long shadow: What led to Hafiz Saeed's arrest.* Retrieved from: https://herald.dawn.com/news/1153708

Fazil, M. D. (2017, March 8). Why India Must Refrain From a Water War With Pakistan: Threatening Pakistan's water supply will have a negative impact on India and all of South Asia. *The Diplomat.* Retrieved from: https://thediplomat.com/2017/03/why-india-must-refrain-from-a-water-war-with-pakistan/

Garver, J. (2001). *Protracted Contest: Sino-Indian Rivalry in the Twentieth Century.* Seattle, WA: University of Washington Press.

Gollnow, S. (2016). *Better Water, Better Jobs – Envisioning a Sustainable Pakistan*. Retrieved from: https://sdpi.org/publications/files/Better-Water_Better-Jobs_Envisioning-a-Sustainable-Pakistan.pdf

Hali, S. M. (2004). *The Baglihar imbroglio*. Retrieved from: http://www.infopak.gov.pk

He, D., Wu, R., Feng, Y., Li, Y., Ding, C., Wang, W., & Yu, D. W. (2014). REVIEW: China's transboundary waters: new paradigms for water and ecological security through applied ecology. *Journal of Applied Ecology*, *51*(5), 1159–1168. doi:10.1111/1365-2664.12298 PMID:25558084

Herald, D. (2005). *Wullar Barrage: Indo-Pak dialogue tomorrow*. Retrieved from: http://www.deccanherald.com/deccanherald/jun272005/national213051200 5626.asp

Hill, D. (2012). Alternative Institutional Arrangements: Managing Transboundary Water Resources in South Asia. *Harvard Asia Quarterly, 14*(3). Retrieved from: http://southasiainstitute.harvard.edu/website/wp-content/uploads/2012/10/HAQ-14.31.pdf

Ho, S. (2017). Power Asymmetry in the China-India Brahmaputra River Dispute. *Asia Pacific Bulletin, 371*. Retrieved from: https://www.eastwestcenter.org/system/tdf/private/apb371.pdf?file=1&type=node&id=35993

Huang, Z. (2015). Case Study on the Water Management of Yaluzangbu/Brahmaputra River. *Georgetown International Environmental Review, 27*(2), 229. Retrieved from: https://gielr.files.wordpress.com/2015/04/huang-final-pdf-27-2.pdf

Immerzeel, W. W., van Beek, L. P., & Bierkens, M. F. (2010). Climate change will affect the Asian water towers. *Science, 328*(5984), 1382–1385. doi:10.1126cience.1183188 PMID:20538947

Iqbal, A. R. (2010). Water Shortage In Pakistan – A Crisis Around The Corner. *ISSRA Papers, 2*(2), 1-13. Retrieved from: http://www.ndu.edu.pk/issra/issra_pub/articles/issra- paper/ISSRA_Papers_Vol2_IssueII_2010/01-Water-Shortage-in-Pakistan-Abdul-Rauf-Iqbal.pdf

Jamwal, A. (2013). *River water Interests/disputes with India's Neighbours as Potential Flash Points*. Institute of Chinese Studies. Retrieved from: http://www.icsin.org/uploads/2015/04/16/647c4123483c6769eceb7b7a10eacf5a.pdf

Jang. (2017, July 30). *Sind Taas Muahida per Pakistan aur Almi Bank k darmiyan baat cheet ka dusra round kal Washington me hoga*. Academic Press.

Jianxue, L. (2010). *Water Security Cooperation and China-India Interactions*. Shui ziyuan anquan hezuo yu ZhongYin guanxi de hudong.

Keetelaar, J. C. (2007). *Transboundary Water Issues in South Asia*. Retrieved on July 20, 2017 from: https://www.google.com.pk/url?sa=t&rct=j&q=&esrc=s&source=web&cd=2&cad=rja&uact=8&ved=0ah UKEwiAxYKjtZ_VAhXIchQKHWGADPcQFgg7MAE&url=https%3A%2F%2Fthesis.eur.nl%2Fpub%2 F4030%2 FFinal%2520version%2520thesis_Jessica%2520Keetelaar_11_03_07.pdf&usg= AFQjCNGX6FHAroi7Tux_qr729P8T5E54OQ

Khadka, N. S. (2016, December 22). Are India and Pakistan set for water wars? *BBC News*. Retrieved from: http://www.bbc.com/news/world-asia-37521897

Khadka, N. S. (2017, September 18). China and India water 'dispute' after border stand-off. *BBC News*. Retrieved from: http://www.bbc.com/news/world-asia-41303082

Khan, M. Z. (2017, January 21). India asked to stop work on Kishanganga and Ratle projects. *Dawn*. Retrieved from: https://www.dawn.com/news/1309767

Klare, M. T. (2002). *Resource Wars: The New Land scape of Global Conflict*. New York, NY: Henry Holt and Company.

Label, L., Xu, J., Bastakoti, R. C., & Lamba, A. (2010). Pursuits of adaptiveness in the shared rivers of Monsoon Asia. *International Environmental Agreement: Politics, Law and Economics*, *10*, 355–375. doi:10.100710784-010-9141-7

Li, L. (2013). An Exploration of the Maturation of Sino-Indian Relations and Its Causes [ZhongYinguanxi zouxiang chengshu ji qi yuanyin tanxi]. *Contemporary International Relations*, *3*, 49–55.

Lipschutz, R. D. (1997). *Damming Troubled Waters: Conflict over the Danube: 1950-2000*. Paper presented at Environment and Security Conference, Institute of War and Peace Studies, New York, NY.

Mahapatra, S. K., & Ratha, K. C. (2016). Brahmaputra River: A bone of contention between India and China. *Water Utility Journal, 13*, 91-99. Retrieved from: http://www.ewra.net/wuj/pdf/WUJ_2016_13_08.pdf

Mirza, M. N. (2016). *Indus water disputes and India-Pakistan relations* (Doctoral dissertation). Retrieved from: https://archiv.ub.uni- heidelberg.de/volltextserver/20915/1/Mirza%20PhD%20Dissertation%20for%20heiDOK.pdf

Mustafa, K. (2014, January 21). India plans new dam on Chenab violating Indus water treaty. *The News*. Retrieved from: http://www.thenews.com.pk/Todays-News-2-227667-India-plans-new-dam-on- Chenab-violating-Indus-water-treaty

Naz, F. (2014). Water: a cause of power politics in South Asia. *Water and Society II. WIT Transactions on Ecology and The Environment, 178*. Retrieved from: https://www.witpress.com/Secure/elibrary/papers/WS13/WS13009FU1.pdf

Pak, J. H. (2016). Challenges in Asia: China, India, and War over Water. *Parameters, 46*(2). Retrieved from http://ssi.armywarcollege.edu/pubs/parameters/issues/Summer_2016/8_Pak.pdf

Parsai, G. (2010, June 1). India, Pakistan resolve Baglihar dam issue. *The Hindu*. Retrieved from: http://www.thehindu.com/news/India-Pakistan-resolve-Baglihar-dam-issue/article16240199.ece

Rediff. (2005). *Baglihar verdict to be binding on Pak, India: Swiss Expert*. Retrieved from: http://www.rediff.com/news/2005/oct/04baglihar.htm

Samaranayake, N., Limaye, S., & Wuthnow, J. (2016). *Water Resource Competition in the Brahmaputra River Basin: China, India, and Bangladesh*. Washington, DC: CNA Analysis and Solution. Retrieved from https://www.cna.org/cna_files/pdf/cna-brahmaputra-study-2016.pdf

Sharma, S. P. (2006). Indo-Pak pact on waters hurdle in power projects. *Tribune News Service*. Retrieved from: http://www.tribuneindia.com/2005/20051008/j&k.htm#1

The Express Tribune. (2016, October 1). *China blocks Brahmaputra River as India threatens to scrap Indus Water Treaty*. Retrieved from: https://tribune.com.pk/story/1191953/china-blocks- brahmaputra-river-india-threatens-scrap-indus-water-treaty/

The Express Tribune. (2017, February 26). *India fails to substantiate claim of Pakistan's involvement in Uri attack*. Retrieved from: https://tribune.com.pk/story/1339596/india-fails-substantiate- claim-pakistans-involvement-uri-attack/

The Nation. (2016, October 17). *India's suicidal water pursuits*. Retrieved from: http://nation.com.pk/17-Oct-2016/india-s-suicidal-water-pursuits

The News. (2016, October 2). *China stops water of Brahmaputra River tributary*. Retrieved from: https://www.thenews.com.pk/print/154331-China-stops-water-of-Brahmaputra-River-tributary

The News. (2017, June 10). *India completes Kishanganga hydropower project without resolving differences*. Retrieved from: https://www.thenews.com.pk/print/155112-India-completes-Kishanganga- hydropower-project-without-resolving-differences

The Times of India. (2016, April 2). *Pakistan claims India 'failed' to provide evidence on Pathankot attack: Report*. Retrieved from: https://timesofindia.indiatimes.com/india/Pakistan-claims-

India-failed-to-provide-evidence-on-Pathankot-attack-Report/articleshow/51665009.cms

Today, P. (2017, February 3). *UNDP report on water security: A wake up call*. Retrieved from: https://www.pakistantoday.com.pk/2017/02/03/undp-report-on-water-security/

Waterman, S. (1984). Partition—A Problem in Political Geography. In P. Taylor & J. House (Eds.), *Political Geography*. London: Croom Helm.

Wirsing, R. (2008). The Kahsmir territorial dispute: The Indus runs through it. *The Brown Journal of World Affairs*, *15*, 225–240.

World Bank. (n.d.). *The Indus Waters Treaty (IWT)*. Retrieved on July 23, 2017 from: http://siteresources.worldbank.org/INTSOUTHASIA/Resources/223497-1105737253588/IndusWatersTreaty1960.pdf

Zawahri, N. (2009). Third party mediation of international river disputes: Lessons from the Indus river. *International Negotiation*, *14*(2), 281–310. doi:10.1163/157180609X432833

Zhang, H. (2015). *China-India Water Disputes: Two Major Misperceptions Revisited*. S. Rajaratnam School of International Studies, Nanyang Technological University, 015. Retrieved from: https://www.rsis.edu.sg/wp-content/uploads/2015/01/CO15015.pdf

Zhang, H. (2016). Sino-Indian water disputes: the coming water wars? *Wiley Interdisciplinary Reviews: Water WIREs Water, 3*, 155–166. Retrieved from: http://onlinelibrary.wiley.com/doi/10.1002/wat2.1123/pdf

Zhifei, L. (2013). Water Security Issues in Sino-Indian Territorial Disputes [ZhongYin lingtu zhengduan zhong de shui ziyuan anquan wenti]. *South Asian Studies Quarterly*, *4*, 29–34.

KEY TERMS AND DEFINITIONS

Arid Agriculture: Agriculture in arid conditions where arid region is characterized by severe lack of water that prevents the growth and development of plants.

Indus Waters Treaty (IWT): Indus Waters Treaty was signed by India and Pakistan in 1960. It was facilitated by the World Bank. According to IWT, the three Western Rivers of the subcontinent, Indus, Jhelum, and Chenab, were given unrestricted availability to Pakistan whereas India was given the unrestricted availability of the three Eastern Rivers, Ravi, Sutlej, and Bias.

International Court of Arbitration (ICA): International Court of Arbitration (ICA) is established in accordance with the Clause (5) of Atricle IX of the IWT.

Permanent Indus Commission (PIWC): The Permanent Indus Commission (PIWC) has been created according to the provisions of the IWT. It provides a contemporary mechanism for conflict resolution and consultation using, inspection, exchange of data, and visits between both the countries.

Tension: Conflict.

Water Scarcity: The shortage or insufficient availability of water.

Water Security: The reliable availability of sufficient water for domestic use. In this chapter, water security has been discussed as an international issue.

Section 3
International Business and Consumer Behavior

Chapter 4
International Businesses and Environmental Issues

Fatma Ince
Mersin University, Turkey

ABSTRACT

This chapter on international businesses and environmental issues addresses the relationship between activities of international businesses and environmental goals. Because of the increasing awareness about environmental issues, the related groups, international policies, and relation force the businesses to be more environmentally friendly and consider the future of resources in their operations. The international regulations, declarations, and other pressures can change the usages of the businesses. The combination of the planet, profit, and people are conceived to protect nature with the aim of growth. From this viewpoint, this chapter provides an overview of the environmental issues, international politics, relations, standards, and successful examples of businesses about being environmentally conscious.

INTRODUCTION

International relations, policies and standards force the businesses to consider the global environmental issues which include eutrophication, global warming, mine-water pollution, air pollution, agricultural pollution, acid precipitation, and deforestation. International committees and other collaborations gather different countries to provide environmental sustainability and reduce the wastes and emission footprint. The acceptance of the environmental approach show differences from one business to another, but the primary goal has the same character for whole institutions as the reducing the impacts of human on the environment. For that purpose, the

DOI: 10.4018/978-1-5225-3990-2.ch004

includes both the evaluating and reducing the firm's environmental impact. Third and the last dimension of the Scheme is transparency, which refers to the necessity of informing the internal and external related groups about environmental involvement (ec.europa.eu, 2017). Also, the public must be informed about the environmental performance of the organizations owing to the transparency principle.

Providing an achieving sustainable improvement and being an environmental leader, EMAS gives the business ten steps and four critical factors. According to the European Commission, the process starts with the contacting the component body and an initial environmental review. Secondly, the Deming Quality cycle constructs the next four steps as do, check, act and plan stages. After this period, the phases of verification and validation, registration, promote exist for using of external processes. These steps are designed to help, improvement, evaluation, impact reduction, reputation, environmental policy, culture change, pollution prevention, legal compliance, corporate responsibility and resource management. At the beginning of the process, it can be seen as a very difficult and complicated Scheme but, after the system starts to adopt the specific benefits emerge. The first advantage of the Scheme is about image management of the business and gaining the enhanced credibility and recognition. External audit and launching are more impressive to the customers and stakeholders. Consumers feel safe with the product of the firm due to the EMAS informing and transparency. The second benefit of the Scheme is Registration List which is followed by the business to achieve the EMAS process. This registration list also has a guiding role in the environmental and organizational requirements. Fulfillment of needs makes the process clear for the organization. Last but not least benefit from the Scheme is the EMAS logo, which leaves the environmental impression on the external stakeholders and provides brand's reliability for the investors and consumers (nqa.com, 2017). And therefore it can be said that EMAS includes the publicly available environmental statement and verified results for the institutions. Owing to the ten steps of the Scheme, EMAS may take a long time, but it is a potent tool for managing both for external and internal shareholders.

International Politics and the Environment

International environmental politics and their intergovernmental aspects are a necessity for the dealing with the global environmental issues. Because the determining the causes of the problems and solutions about them, the term requires a multinational and integrated approaches and rules. Human impact on the environment needed international attention, awareness and efforts are the concepts of the global environmental politics. These politics is not only the cooperation and collaboration among governments or institutes, but it is also about the conflict between all parties concerned due to the environmental problems. Environmental degradation and natural resource use or human-generated impacts may create a conflict of interest (Mitchell, 2010). The environment is changed by humans deliberately or unconsciously for many centuries. These effects, which were seen in the early ages, increased further with industrialization and it became uncontrollable. From an agricultural revolution to the industrial revolution, there are various human impacts seen as a significant environmental disaster. The change process starting with the discovery of tools continues the present with reducing natural resources. Industrial revolution promotes the growing population and using the animals and other resources of the planet for the humans.

Due to the industrialization, the increasing amount of the environmental accidents or problems such as eutrophication, mine water pollution, air pollution, global warming, acid precipitation, and deforestation requires the common and acceptable concern being an essential part of foreign policy and international relations. Reducing the human effects on the environment can be real with the self-conscious efforts of people, companies, and countries. Opponents, supporters and other bystanders have responsibilities as well as enterprises. To prompt all interested parties and to prevent conflicts that will emerge in this process, the study of the international environmental politics should be the leader, which include the reason and the concept of the conflicts arise over environmental issues and the solution methods. Additionally, the failed resolutions and its reasons are also the subjects of these politics because of the giving information about when the negotiations get nowhere.

International environmental politics use a multidisciplinary approach which involves law, economy, chemistry, biology and philosophy modeling notwithstanding that political science to understand the environmental issues. Therefore, assuming the adoption of protocols accepted different countries have different costs for the interested groups. Environmental politics provide the sharing cost among parties to ensure the prevent the specific pollution. However, the protection the wild nature, other natural resources, and species need the international scene to avoid being a global wasteland. The first known example of the environmental policy is the Panel on Climate Change and United Nations' Conference on the Human Environment (UNCHE) in Stockholm 1972. There is also a lot of study of international environmental politics, but the impact of them are not always as expected, so they cannot become widespread. Another popular development is the UN Conference on Environment and Development (UNCED) in Rio de Janerio in 1992 (Najam, 2005). The results of the conference take part in two essential journals to generalize the outcomes and arouse interest. After these critical meetings or conferences, the combination of trade and environment become the main topic of international environmental politics. Nowadays, studies are continuing in the light of the sustainability.

International Relations and the Environment

Environmental issues indirectly and directly affect different countries at the same time. Inherently, the pollutions and scarce sources require international diplomacy and relations. After the Stockholm and Rio Conferences, international environmental ties draw high interest. Both Stockholm and Rio Declarations involve some different principles. Especially the developing countries are not willing to pay the environmental costs with the new environmental programs. Similarly, after the Rio Conference 180 countries and private organizations which try to influence policymakers don't want to accept all subjects, and then only 27 principles are signed.

In the light of these developments, international environmental relations are developed to avoid the long-term environmental issues and global press about them. Climate change, marine resources, and diversity of species are some of the subjects of international environmental relations as well as hazardous wastes, nuclear pollution, and environmental security. The main target of the countries known as sustainable development should occur with environmental consciousness. As the report on Our Common Future emphasized in 1987 by the Brundtland Commission, the environment must be a part of the development and poverty (Harris, 2002). Therefore, the environment should not be considered separately from the sustainability and development of the countries.

From this point forth, the environment and international relations involve the study of the global environment and theoretical approaches used in international relations. The environment in international relations. International ties interest social ecology and the globalization of the environmental changes or issues as well as safeguarding the planet. First international relation studies are mostly about conflicts, insecurity and possible cooperations. After the Great War and changed aspect, optimistic and liberal reforming project occur in the international system to build collaborations. Environmental committees and their conferences guide the countries to decide their stand against the global environmental issues. Maritime and air pollution takes place in the international relationship framework mostly owing to the Antarctic Treaty regime (Vogler & Imber, 2005). Additionally, the impacts of human are mainly analyzed to understand the responsibility of the issues. The common point of all studies is needed to the cooperation between countries and agencies.

UN, WTO, and OECD as the major international institutions are developed for the economic growth and global trade. Similarly, environmental international relations require the committees which can solve the problems of growth and the environment. At this stage, it is so important to understand the environmental-based growth and its effects on the other activities. Inherently, international relations have a holistic approach to the environmental and economic issues among governments (Purdey, 2010). The evidence and the impacts on the environment are known by the interest groups in a short period. But, international institutions and other economic and politic policymaking cannot avoid the growth paradigm in the global structure. Also, the long-term chain reaction effect of environmental issues is hardly measurable for the earth. Because the results of a natural disaster in one country may be the beginning of a problem in another country. Herewith, environmental issues cannot be considered as a unique problem in society. So, the nations should integrate the economic, environmental and social targets with the international relations and policies. This is an unavoidable consequence of the international ties between private and public institutions.

ENVIRONMENTAL ISSUES

Global warming, air pollution, mine and water pollution, deforestation, acid precipitation, agricultural pollution, eutrophication are the major environmental issues. Therefore, these global issues are explained in this part of the chapter because of the importance of the subject.

Global Warming

It is one of the most known environmental matters nowadays. First measurements of the earth's atmosphere are taken in the 1860s, and now the temperature of the atmosphere is warmer than before. The climate and the dates of the mid-seasons changed. This critical issue is so popular because the people can feel the change in their daily life. The planet can live and save it habitable characteristics at this level but, the greenhouse effect is getting more and more, so the medium level will be passed in the future. Carbon dioxide, nitrogen, oxygen, and other gases in the atmosphere above the land has a delicate balance. And this chemical balance protects the earth from the short and longwave radiation of the sun and provide a suitable warmth for the habitat and life. The greenhouse effect damages the layers of gases, which absorb the short and long wave radiation and it causes the radiation to pass through and warm the air and surface (Best, 1999). Especially the amount of carbon dioxide which increases due to the human activities such as using fossil fuels, growing population, is getting more and more.

The seriousness of the problem can be realized after some diseases and political stages. International bureaucracies raise awareness about global temperatures through Panel on Climate Change (IPCC) in 1990. Two years later, about 154 countries come together at the Earth Summit in Rio de Janerio, Brazil to decrease the emission with the voluntary agreement. But an optional step is not enough to meet the needs. So, there is need a specific target for the meeting of the emissions level. In 1997, Kyoto protocol signed to bind countries to emission reductions, because the effects of greenhouse gases can be felt in the social and economic life. The greenhouse effect on the earth warming is agreed by the scientist, after the Svante Arrhenius article in 1896. The author points out that humans are enhancing the damage of the warming and increase the natural greenhouse effect. After the observing the warm and cold period of the earth, this theory is promoted by the several scientists (Jones, 1997).

In the twenty-first century, the impact of climate change, global warming or greenhouse effect is estimated with various aspects. A complex network of evolution shows that of coastal areas for agriculture, there are a lot of direct and indirect effects exist in the ecosystems on earth. Thermal expansion of the oceans and the rising of the sea level will affect the coastal areas seriously. The freshwater resources can damage because of the unbalanced weather events and climate changed. According to the lack of water and acid rain, the food supply will decrease staidly and may risk people's life. Sustainable agriculture is getting hard and costly in a short period. Also, the quality of the air will not be enough for the plant and peoples' life. The impact of human health has already started in some countries such as Africa and China. Back to nature and fix the balance cannot be possible after the critical level of warming but, to this level, any applications or industrial implementation can be seen in the adaptation of the climate change.

Air Pollution

Air quality has vital importance for the human and natural life. Especially big cities have unhealthy conditions for breath because of the cumulative bonfire or coal fire of homes. In the early days, Henry III and Queen Eleanor leave the castle because of the coal fires of the houses of the kingdom and choose another place in the fresh air. After the modern businesses, Parliament starts the Public Health Act to decrease the air pollution of factories in 1936. Some committees have appointed, and various steps were taken, but the world meets the air pollution and its results in London events dramatically. In 1952, great and terrible smog descended in London for five days. The combination of the warm and cold air lead such a smoke and moisture, fog and it blankets the area suddenly. Four thousand people died because of the breathing difficulties, crash and other events as a result of the smog and bad weather quality (Best, 1999). This tristful event takes the world's attention on the incident and the importance of the air quality.

The impact of the London smog is profound, and The Clean Air Act starts for some control areas. The air plan is moved into action to decrease the amount of SO_2 and change the domestic fuel. In 1990 the emission level of the UK falls from 6,3 million tons to 3,7 million tons. This considerable reduction provides an improved air and rainwater quality and helps to minimize the acid degree of rains. The importance of the air quality requires multinational solutions and collective steps. Because the air pollution of the China can affect the USA in an extended period and the deterioration of the atmosphere has a common characteristic for all nations or countries. After the UK legislation, EU and UNECE start air-quality legislation in the member countries. Also, World Health Organization, Asia legislations, EU industrial emissions legislation and vehicle emissions exist (Tiwary & Colls, 2010). All standards and legislation serve the main aim of the controlling the air pollutions and its harmful effects. In short and long-term effects of air quality on human and other species has a survival extent. Also, the sources and control of gases are about the lifestyle of the modern society to a large extent.

Air pollution should be handled different aspects because of the complicated balance of the atmosphere. The primary and vertical structure of the atmosphere has a critical importance for absorption of gases and minimize the other air pollutants. Plants, humans, and animals have the effects of the combined of the gases and aerosol. Pollution is named anthropogenic, which arises from human activities, while it is called biogenic which I about animals and plants (Popescu & Lonel, 2010). Because the air pollution is handled in connection with environmental damage and its specific capacity to damage life system with toxins and ammonia or carbon dioxide contents. The changeover of the origin of the atmosphere has normal and abnormal periods, but some regular disturbances change the natural constituents of air permanently and detrimentally.

Mine and Water Pollution

It is widely known that the planet's surface consists of 70 percent water and mostly it cannot be used by humans because of the ice caps at the Poles. Humans can use the only freshwater which is the 30 percent of water and on the surface. Rivers, lakes, and wetlands provide fresh and usable water from underground to the surface. Not only for the drink, but also for irrigating crops there needs a certain amount of clean water. Also, power mills and machines or some industries require water for cooling and force. Remaining from drinking water is used for sewage disposal. Growing population also uses more fresh water day after day. Increased demand for water is met with drained water pumped from underground (Fridell, 2006). Therefore, lakes, rivers, and barrages cannot be fed from underground and open mines used to be.

Water pollution also arises from a lot of reasons such as oil pollution, sewage pollution, organic and inorganic pollutions. Thermal pollution is also classified in this kind of pollution or disaster. So, water pollution is emerging as a threat to all creatives directly and indirectly. The quality of the water is about to the chemical, biological and even physical content of the water which includes not only organic and inorganic substances in solution but also identity, temperature and impact or organism in it. The atmosphere, rocks, soil shape the water characteristic with cultivation practices or vegetation canopy. Therefore, the pollution of the water originates in different resources such as natural process, agriculture as well as deforestation, urbanization, and chemicals. Solid or liquid-solid washes from industry, dwelling areas, disposal of gaseous (Agarwal, 2009). On the other hands, the effects of the industrial wastes or chemicals have the damage power of nature and human life. Uncontrolled human activities are the most effective response to the accelerated flow of minerals which form the water and its quality.

Radiologically or biologically, the substances of water lose quality and become usefulness form of nature. Moreover, discharges of wastes include some radioactive effects and can change the temperature, taste or odor of the water. In addition to that, any impairment of water quality is seen as water pollution because it has the unsuitable form for beneficial use hereafter. Natural reasons for the decreasing quality of the water are not named water pollution such as atmospheric dissolved gases, decomposition of animal or vegetable materials. These effects are called non-point sources while others are known as point sources such as individual and industrial sources which include agricultural or municipal waste, spillage of oil and watercraft waste. Sand, paper, sewage, foam, and pigments are the physical reasons for the dissolution of the water, while fluorides, salts, phosphates, detergents, plastic are the organic and inorganic chemical reasons of the pollution. Additionally, bacteria, nematodes, protozoans, slime, and nematodes can damage humans, or animal health is known as biological factors of the water pollution (Abbasi & Abbasi, 2012). Hence, the pollution of the water has various properties and struggle with it is needed the systematic aspects of different science and foundations.

Deforestation

Ecological well-being is affected by global deforestation and its impact on the earth. 10 percent of the earth is covered by forest, while 20 percent of it is the continental area. Developing or tropical countries have more forest area than developed states with 55 percent part. Canada, China, Democratic Republic of Congo, Brazil, Russia, Indonesia, and the USA have more than 60 percent of the whole forests which is the considerable part of the ecosystem. Accordingly, these forests provide a useful living space and the most extensive habitat for animals and other creatures. Forests, which supply fresh air also can stop soil erosion and produce wood or fuel for energy (Vajpeyi, 2001). Additionally, plants which grow in the forest are used to the medicine and help people food control for the sustainability of the economic and ecological system. Especially the tropical forest cover is changing over time due to using the trees for industrial consumption.

Taking into account all of these, replanting is not enough and the forests disappearing dramatically. To a level, the forest has some reproductive and restorative features due to the evolutionary diversity and genetic traits. After the renewable level, destruction is beginning and then diversity decrease under the risk level. Modern-day efficient like new roads and buildings expand the rate of deforestation as well as industrialization. Extreme climatic conditions and insect pests are other reasons for the deforestation and disappearing the forests swiftly. Deforestation makes major differences in the environment through the impact of destructive and manifold (Kaimowits & Angelsen, 1998). As a result of all this process has the adverse effect of the sinking of carbon and absorbing carbon dioxide. It is not only about carbon, but this process also causes the warming of climate because of the unequal water recycling and then climate warming. The knock-on effect of hydrology triggers the recession of glaciers and uneven flows due to decreasing of flow in rivers.

Margulis (2004) explains the deforestation in a broad aspect and take the subject in hand with geologically, ecologically and economically. Because governments feel pressure about opening more roads and serving large masses due to the population increase. As the society develops, the needs grow, and there are some new demands which affect the geo-ecologic conditions and environment. Deforestation for the new roads is one of these claims and pressure. Also, new buildings and roads have geopolitical subjects of the countries, as well as them, give profitable chances for the firms. So, it can be said that a balance must be established between profitability, costs and social benefits for the future of the forest and nature. Economic activities and targets as the aim of the firms cannot be the only dimension of the economy. Because protection of the forest as a social target is not only a non-profit target but also about the future of the economy and human life. The results of the deforestation have a relationship with economic factors as well as social cost and benefits. Therefore, deforestation has multinational boundaries and various dimensions for the sustainability of the natural resources.

Acid Precipitation

Acid or rain precipitation is about the unremovable airborne pollutants in rain droplets. Acidification of snow or rain can be seen on the effects of buildings, plants, and trees as well as human health. First damage from acid rain is seen as a local problem and result with the close the chemical factories. Widespread dispersion of the pollutants cannot be seen in early times. After the growing activities of the heavy industries and energy firms, the damage to the atmosphere shows an excessive increase. The waste gases from heavy industries or energy firms provide the acidic gases into nature, such as sulfur dioxide, nitric oxide, nitrous oxide and nitrogen dioxide. Road traffic and other human activities help the increasing damage of the gases and its chemical reactions. These gases undergo chemical exchange with sunlight and cloud water and become more acidic for some countries such as Norwegians and Swedes (Best, 1999). Also, the problems of the extreme snowmelt and black snow which is about the acidic and dirty snowflake are not a local issue. However, black snow and unbalanced pH level of meltwater kill the aquatic organisms and fishes. Thus, the problem of acidity is both about pH level and other chemical compounds of water.

In addition to water life, acid precipitation becomes a threat not only to the forests and agriculture but also for human health and sustainability of the earth. The harmful effect of acidic raindrops on nonindustrial areas is seen more than industrial areas due to the moving around of gases. Gases and their particles are lighter than air, so they can carry themselves long distances and cause the environmental loss in wild areas, while smoke particles are heavier than air and they sink to the area which is near the factories. When the some reach the ground level, it makes breathing problems for a living creature because of the feature of thinning and range over a wide field. Prevailing winds are the main reason of the polluting gases spread other areas at high speeds (Petheram, 2002). Unsurprisingly, the acidic water or current air damage, the sources of the neighborhood countries. But, surprisingly the damage can reach the other continents with wind and flow. For this reason, the issue of acid precipitation is a universal issue, at least as much as other environmental problems.

According to the Heij and Erisman (1995), acid deposition has large-scale impacts of forest and forest soils. Therefore the measurement of the ecological effects of the atmospheric deposition on forest and non-forest ecosystems are so difficult and complicated. The research about acidic contents of the rain or other water sources is shown that there is the need to integrate acidic deposition modeling and solves with supra-national aspects. Developed countries as Canada, US, and Switzerland speed up the study about the causes and effects of the acid precipitation as well as trying the solve this universal problem for the next generations (Lane, 2003). Because the acid deposition can endanger the areas in the wet and dry form which include acidic rain, toxic fog, black snow and acidic particles or gases.

Agricultural Pollution

Pollution from farming has various effects on nature which are both about animals and humans. Especially monoculture farming treat the animals and human health unexpectedly as well as the effect on the sustainability. This effect takes place through feedlots which are seen as a super productive and effective tool of the meat-eating demands. Feedlots and monoculture are both harmful effects on the soil and water chemistry. Because the waste of animals in a small space makes soiled the natural resources. Some regulations are sanctioned by the government to decrease the pollution, but they are not enough for nature. Disastrous results have still seen in the industries of the food, agriculture, and livestock (Fridell, 2006). The immediate influence of this pollution can be seen due to consuming the grain, pesticides, drugs, petroleum, water, and other natural sources.

Key routes of the agriculture pollute the environment, classify into two sections as contamination of water and food. Fertilizers, nitrates, and pesticides which are used in the agricultural productivity can cause the cancers, taste problems and other dreadful illness through water consumers, fisheries, tourism, surface water transport as well as agriculture sector. Also, organic livestock manures make the water plus deoxygenation, while soil leads the filling of lakes or harbor and destroying coral reef and fisheries. Additionally, ground and surface waters wash the pesticides away and affect water's quality mortally. The wildlife, agriculture and other industries are also affected by the livestock manures, paddy rice and biomass burning through food consumers, farmers, other users. Methane which is seen as a result of paddy rice and ruminant livestock has a critical role in global warming and acid rain (Angell, Comer & Wilkinson, 1990). This gases with other smoke and particulates can damage the health seriously and irreversibly.

Wildlife and landscape protection must be one of the main aims of the Agri-environmental policies to decrease the phosphate and nitrate pollution of lakes, rivers and streams as well as the spillages of silage effluent from storage areas. Also, to avoid pesticide drift and to control the accidental spillages of diesel oil from tanks, there is a need for some regulations embracing the agricultural sector. In addition to all these, runoff water from vegetable washing plants is considered as a kind of pollution from farming. Unfortunately, soil erosion is not recognized as a serious problem as well as dairy farming areas (Shortle & Abler, 2001). Water pollution due to farming is not just about the fish-killing; it is a severe problem with a wide exhaustive. Using of manure on frozen fields, spraying pesticides and other agricultural practices need farmer education and some policies which consider all dimensions of farming pollution with water and air pollution as well as soil conservation.

Eutrophication

The eutrophication threat to dynamics of aquatic ecosystems both in local and global. Tourism and lakes are directly affected, but the changes of the eutrophication balance of the ecosystem. Salt marshes, fresh water, and plankton get also changed crucially due to eutrophication processes. The eutrophication of water is about over-enrichment by phosphate and nitrogen due to the decomposition of algal organic matter. Because the deterioration of the water chemistry leads the oxygen depletion in water bodies. Point and non-point resources of the nutrients consist of industrial discharges, sewage, agricultural and urban outcomes. Especially excessive application of fertilizer cause phosphorus in soils (Ansari & Gill, 2014). Then, these phosphorus-rich soils reach lakes and other water sources and trigger the growth of aquatic plants and phytoplankton. Drivers of the ecosystem always interact with each other in a synergic way. Hence, eutrophication causes more energy loss and increase the consumption of fertilizer owing to direct and indirect drivers.

Harmful algal blooms are common issues in lakes with eutrophication. Strong lake lands, air water or water sediments interactions make the water body available to chemical degradation in shallow lakes. At the end of the eutrophication and algal blooms process, shortage of fresh water or drinking water and forced degradation of lake ecosystems are seen as a big problem for the environment and supply of water. Algal blooms which are also named microcystins is about the deterioration of water quality, and they especially occur in the summer owing to the decreasing water supply. Industrial wastes enhance the degradation with sewage disposal and make the eco-environmental problems bigger in lakes. Lakes have their hydrodynamics and potential for the renewable their water bodies unless intervened the process outside. However, remediation of eutrophication in Aqua ecosystems is the first of the nutrient pollution of sources (Quin, Liu & Havens, 2007). It means that eutrophication is the beginning of the sequence environmental problems in aquatic ecosystems around the world. Water quality and fish population are the first articulable evidence of the disorder.

The problem of the rendering rich in nutrients are considered the last 30 years by the countries to save the fresh water of lakes in the worldwide. Human activities enrich the water with nutrients unintended and change the lake's life cycle or ecosystem permanently. Other factors can also affect the ecosystem of the lakes, but the harmless one is wasted of the human or factories (Somlyody & Straten, 2012). This situation must be remedied to avoid the changing of the original functions of the lakes. A watershed is seen as a solving tool for this problem in the shortest time, while the natural assimilation process is considered as a lasting solution for the sustainability of the lakes. Development of the countries provides through urbanization, agriculture, tourism, and industrial. But, countries are now aware of increasing of nutrition in the lakes and they start to consider the results of the process in the near future. The solving of the problem needs a broad scanning owing to the nature of the lake ecosystem which includes both external natural factors and interrelated chemical, physical, biological process.

Other Important Issues

Environmental issues are of the whole economic, social and politic dimensions of a country. Therefore, one of the important issues about the environment is ceramics and so glasses which are used by the nuclear industry and so important for the defense industry. Because of the importance of the subject, government use regulations to manipulate the energy. The nuclear industry has high-level and low-level radioactive outcomes due to the ceramic or glass. Thus, countries try to make environmental stewardship to growth with nuclear and environmentally friendly products. Ceramic Science and Technology for Nuclear Industry and Science and Technology in Addressing Environmental Issues in the Ceramic Industry are two of the symposia about ceramic and nuclear which are organized by The American Ceramic Society (Smith, Sundaram & Spearing, 2002). This society organizes various activities as well as other foundations and public enterprises. Product stewardship, refractory users, markets and some policies and other factors force corporations and producers consider the environment and recycling of the resources for the nuclear industry.

Other energy resources are classified into three factors as radiant energy emitted by the sun, gravitational energy resulting from the interactions of the Moon, Sun, and Earth, and geothermal power originated inside the planet. Thus, the resources are mainly four kinds of nuclear energy which is abundant yet exhaustible. But, all these resources of energy are not efficiently usable or available to use because of some geological, wind regime and hydrological reasons. Some of them are provided with economically, while others have 50 percent probability of use or 10 percent probability of exploitation under some circumstances. Because the supply of energy is a manner of a geopolitics position of a country and internal reserves determine the international trade and negotiations of environment against foreign or borders (Goldemberg & Lucon, 2010). For this reason, energy resources have a vital importance, and it is a critical issue for the countries. Because it is one of the significant tools of the sustained growth of an economy or country as well as it is about the defense system.

Additionally, some natural hazards can affect the balance of the energy resources and the environment. But the human impact of nature can be more harmful in the long term. Because human has a lot of roles as cultivators, metal workers, and keepers in nature. These roles have effects on vegetation, grassy plants, and forest through grazing and firing. Also, the human has some negative influences on animals and wild nature with the dispersal and domestication of animals. Directly or indirectly, human lead the animal decline and habitat change solicitously (Pickering & Owen, 1997). All these old and recent developments oblige alternative assessments with the environment and new assignments for all special interest groups. After the last events, it has become a necessity, not a precaution for the future.

Global Activities of the Businesses About Environment

There are successful examples of environmentally successful in different sectors. Especially renewable energy sector draws attention as well as cleaning producer. In this section, some of the most exciting and successful firms are given as examples of environmentally friendly companies. However, the journal of Forbes (2017) also provides extensive publicity to these businesses as the best firms for the environment. The importance of these samples arises from the mission of showing other big companies that it is impossible the being environmentally sustainable when still making the profit.

Werner & Mertz known as FROSCH introduce itself as a well established on the European and Japan market as an innovative company which always feel equally committed to the principles of an environmentally sound and sustainable business approach. Because the firm considers the environment that sustainability is not a fad, it is a company tradition for the daily activities. For 150 years, they provide innovative solutions for the consumers and give importance loyalty of employees. The firm sells the products to 15 countries and has an innovative leadership aim in the target market. The business started as a small green eco-pioneer initially, then the objective of the producing environmentally friendly products expanded. After the serious environmental incidents, Germans become to be the high level of awareness about environmental issues; then the firm notices the demands for environmental products. Consumers enjoy the feeling good eco-quality when they are meeting their requirements. Therefore, environmental alternatives become favorite among consumers. Firstly, phosphate-free detergents launch to the market, then the products of environmentally friendly cleaning are expanded. To increase the consumer trust, the firm uses the environmental solutions for the process and activities. Resource efficiency and environment certified (EMAS III) gathered with value-added supply chain, green frog, and waste management. The firm gains a medal as the most trusted brand about the environment, and it provides environmentally friendly cleaning and care products in nearly all areas of the household. Cleaning, dishwashing, washing, baby care, room freshener and creme soap are some of the products of the firm in the market. "Feel good eco-quality since 1986" is the slogan of the company and show it as a successful eco-pioneer with the environmental and skin-friendly properties of the products. The nine aspects of the aim of the creating sustainable products are naturally based ingredients, skin-friendliness, vegetable origin, organic compounds, reducing utilization of packaging materials, vegan, energy-conscious production, experiences, and water treatment plant (Frosch, 2017).

Another successful firm in environmental business is in the same sector. Seventh Generation is a biggest plant-based producer certified as a B Corporation by Vermont which means that the company is better for the environment and employees. To create a good future for the earth, the firm indicates that trying to reduce the environmental impacts of the products with the sustainable supply chain. According to the company, performance and safety should provide the more mindful way of doing business with the customers and stakeholders. This is the critical factor in the making a difference in the target market. The firm produces the household, baby care and feminine products such as diapers, laundry, cleaners, tampons, baby skin care, recycled paper and Botanic disinfectants with the considering the impacts of the decisions on the next seven generations. Since 1988, firm sustainably produce the goods with the participant of the employees. Renewable energy and low emission cars using are promoted while the solution for the sustainability of employees is rewarded. The slogan of the firm is "it's time to tell the industry to come clean" (Seventhgeneration, 2017) and it shows the environmentally friendly aspects of the operations.

Similarly, Seventh Generation which has the top score is a competitor, METHOD also product environmentally friendly personal care and cleaning product such as dish soap, laundry detergent, all-purpose soaps and shower gels. The firm offers the consumers the naturally-derived power to fight stain with the aim of reducing plastic waste and helping to save the environment. The company has a benefit blueprint program to increase the C Corp Score, compass score and to reduce the carbon footprint, water usage.firm has a green glossary for the customers that gave high-level environmental awareness and it uses recyclability evaluated materials in all products to reach the animal and nature friendly aims at reducing carbon emissions, using renewable energies and offsetting the supplier to be environmental (Methodhome, 2017). From the design of the transfer of products, all processes include green sourcing programs, fuel efficiency, and biodiesel fleet.

Other firms that produce environmentally friendly products for nature lovers or campers and is very successful in this section. Go Lite produced various goods as outdoor clothing and equipment maker since 1998. Ultra-light packs, down jackets and sleeping bags are produced with the aim of the mitigating 100% of the firm environmental footprint. For this purpose, company use recycled nylon and polyester as Environmentally Preferred Materials instead of petrochemical raw materials. Firm gain the 67 percent of the minimizing the footprint. The buildings and factories of the company try to prove a zero waste facility with recycling and composting of all wastes (GoLite, 2017). The company provides the nature lovers gears to go the distance. Therefore, it can be said that the company serves the customers with their environmental sensitivity.

One of the popular environmentalist firms is an Ale company.New Belgium Brewing is a maker of Fat Tire Amber Ale since 1991. The founder of the company is a bike tripper, and he encourages the employees to get on the bicycle. The firm reduces 75 percent waste of its products and using of energy; also the managers consider the emissions and recycles as an important gain for the sustainability. Employees purchase the share from the firm, and this provides commitment and awareness of the environmental subjects. The principle of the employees is "hard work and continuous push for the invention." The vision of the firm includes ten core values and beliefs. Two of these values are directly about the environment. First one is "Kindling social, environmental and cultural change as a business role model", while the second one is "Environmental stewardship: Honoring nature at every turn of the business". Innovative, quality and efficiency are trying to provide to the participant of the employees and increasing environmental processes (NewBelgium, 2017). The firm has specific targets for reducing CO_2, climate change, wastes, water use ratio, protect and restore the waterways.

According to the article on the Forbes (2017), the other firm considered better for the environment is in the manufacturing sector. Solberg Manufacturing product for the protection of the machinery and its surroundings with renewable energy and better energy using in the within the scope of the Better Buildings Better Plants program. The firm firstly examines the entire operation and then try to do more sustainable practices about designs, manufacturing, and processes. The slogan of the company as "protect your equipment, protection your environment" serves the environmentally friendly activities. According to this aspect, firm gain "Best for the Environment Award" and take a certified as B Corporation. The view of sustainability starts with top managers and continue the lowest employees in the firm (SolbergMfg, 2017). Thus, corporate and social responsibilities are considered seriously as well as other market targets.

Successful environmentalist firms can be seen in any sector. This company is one of the most important manufacturers of the 70-plus employees. West Paw Design produces eco-friendly dog toys, beds, and mats with fiber from recycled plastic bottles since 1996. However, the firm use recycled paper and soy-based inks for packaging the toys and beds. Also, customers are invited to send back used products for recycling to reduce the resource uses. The story of the firm starts with five animal lover employees, and then 15 employees take part in this group in Asia to avoid outsourcing and promote local talent to the work life. The first aim of the firm sets out the world's safest and highest quality dog and cat toys with the eco-friendly manufacturing facility. Montana gives inspirations to the employee for sustainable practices in manufacturing or designs. There is a Join the Loop program to provide infinitely recycled toys. Therefore, the firm' values give the employees make differences when they produce the fun toys for the cat and dog lovers. Firms gain the Certified B Corp as one of the best small companies and first benefit corporation of the America. Nowadays, only less than 80 employees are designing toys to achieve the aim of being the world's most eco-friendly and safest pet products. Organic catnip and re-usable banana boxes are some of the environmentally friendly products of the West Paw Design. They also eliminate the buying or using plastic totes, and they are the first IntelliLoft user which includes recycled plastic foil and fabrics. Only eco-friendly materials are used for the product such as organically-grown hemp, catnip, cotton and stuffing made from recycled plastic, bottles made from old chew toys (WestPaw, 2017). At the end of the all these design processes, employees feel the happiness of the keeping waste out of landfills, chemicals out of soil and water. They have fun with the design of extraordinarily toys and beds with the satisfaction of being a part of an eco-conscious firm.

The Eco friend product is mostly about the energy demands of the household as well as heavy industry. Therefore, some prospering firms offer an environmentally friendly energy product for the consumers. Sungevity is an energy supply firm which helps household and industry to reduce the energy losses with the renewable natural energy. The slogan of the Sungevity is "we only have one home; it should be powered by sunshine, " which shows the environmental aspect of the firm. The firm provides high-quality solar panel technology for the homes with solar panels known as Sungevity Energy System to reduce the cost of energy for users. The first step the process is the decision of the consumers to power his/her home with sunshine. Solar energy can be used over 20 years, and it gives the people the chance of saving the nature, trees, and atmosphere. Firm show the consumers the amount of the energy using and the level of the damage to nature statistically and clearly (Sungevity, 2017). Due to the sun based energy system, consumers and nature bear less cost for the sources.

Similar to the previous firm Namaste Solar offer environmental, energy systems for the consumers successfully. The firm claim responsibility of impacting on the environment as well as customer and employee satisfaction in its operations and maintenance services. The slogan of the firm is "transforming energy, transforming business" with the target of the contributing the long-term health of the society and the environment. Owing to the shared culture among employees and flexible working arrangement, the firm target the work in a consensus building which allow the ideas are growing and supported creatively. According to the firm' cooperative model, a strong relationship is gained among inner and outer partners and being a good neighbor to the residents (NamasteSolar, 2017). The company gives the opportunity to work at home for employees, and this practice fell them motivated to new solutions. Another successful example of the energy firm is in Mexico, which there is a leading firm in the solar industry and the firm notices with the employee-owned structure and largest certificate of the Benefits Corporation (B Corp). The main target is customer satisfaction with low environmental impact and leading residential and commercial solar installer by Positive Energy Solar. The name of the firm is SunPower, and it runs the business with the principles of combining People, Planet, and Profit on the same importance level (PositiveEnergySolar, 2017). However, the firm prefers local labor force, employment to improve the local communities it operates and lives in.

To solve the problem with next-generation waste reduction programs and American firm is running with certified B-Corp. The company aims to cut the trash in half across the United States by the name of WasteZero. Managing the wastes have financial and environmental costs for both private and public sectors. Since 1991, the firm has helped the cities, counties, state agencies or private organizations reduce costs and waste when increasing recycling. At first, the company analyzes a municipality or company's waste situation, and then design a powerful waste reduction program for the consumers. After the detailed analyses, the new program is built and launched with a management and control system. At the end of the process, the firm helps the commenting the results and comparing with the modernization program (WasteZero, 2017). The concept of this corporation is different others owing to the providing support services for various companies.

As it is seen clearly from the examples, solar energy or green energy and detergent firms have successful environmental products and services as well as recycling firms. The common point of these businesses is the synergy, structure of the work system which combined environmental and employee friendly processes. They value both people and planet together, while they gain the financial targets successfully and sustainably.

CONCLUSION

In the third title of the chapter, the successful and liked businesses are seen with their positive image on the public. From the stage of the product design to the transporting to the consumers, the business can act an environmentally sensitive and operate the process with environmentally. Because the second title of the chapter shows the environmental issues reach the serious level for all interested groups and after a while, it may be too late for everything. As seen the first title of the chapter, since the 1970s there various environmental steps in the international scope. Unfortunately, all meetings, conferences or declarations don't gain the environmental targets. However, the standards and international committees force the businesses being innovative about the environment and try to show the advantages of environmental consciousness. Butterfly effect can be seen globally in the environmental issues. Inherently, air pollution triggers the global warming and acid rain, while deforestation causes eutrophication and decreasing the air quality. And therefore, all issues are considered as the multidisciplinary and integrated approach to international relations contribution. The attitude about international policies and common goals are so important for the future of the natural resources. The proactive approaches provide a holistic view of the business instead of cleaning sewerages or wastes management.

REFERENCES

Abbasi, T., & Abbasi, S. A. (2012). *Water Quality Indices*. Amsterdam: Elsevier. doi:10.1016/B978-0-444-54304-2.00016-6

Angell, D. J., Comer, J. D., & Wilkinson, M. L. (Eds.). (1990). *Sustaining Earth: response to the environmental threat*. London: Macmillan. doi:10.1007/978-1-349-21091-6

Ansari, A. A., & Gill, S. S. (2014). *Eutrophication: causes, consequences and control* (Vol. 2). London: Springer. doi:10.1007/978-94-007-7814-6

Best, G. (1999). *Environmental Pollution Studies*. Liverpool: Liverpool University Press. doi:10.5949/UPO9781846313035

EPA. (2007). *Learn About Environmental Management Systems*. Retrieved July 19, 2017, from https: //www.epa. gov/ems /learn-about-environmental- management-systems

Forbes. (2017). *11 Companies Considered Best For The Environment*. Retrieved July 18, 2017, from https:/ /www.forbes.com/sites/susanadams /2014/04/22/11-companies- considered- best-for-the-environment/#78b5119212ae

Fridell, R. (2006). *Environmental Issues*. Marshallcavendish.

Frosch. (2017). *Eco-quality in 9 aspects*. Retrieved July 18, 2017, from http:/ /www.frosch.de /Brand/Eco-quality-in-9-aspects

Garcia, S. M. (2003). *The ecosystem approach to fisheries: issues, terminology, principles, institutional foundations, implementation and outlook (No. 443)*. Rome: Food & Agriculture Org.

Goldemberg, J., & Lucon, O. (2010). *Energy, environment and development* (2nd ed.). London: Earthscan Publications Ltd.

GoLite. (2017). *Gear to Go the Distance*. Retrieved July 18, 2017, from http://www. golite.com/

Harris, P. G. (Ed.). (2002). *The environment, international relations, and US foreign policy*. Washington, DC: Georgetown University Press.

Houghton, J. (2009). *Global warming: the complete briefing* (4th ed.). Cambridge, UK: Cambridge University Press. doi:10.1017/CBO9780511841590

Imber, M., & Vogler, J. (Eds.). (2005). *Environment and International Relations*. London: Routledge.

Jones, L. (1997). *Global Warming: The Science and the Politics*. The Fraser Institute.

Lane, C. N. (2003). *Acid rain: overview and abstracts*. New York: Nova Science Publishers.

Margulis, S. (2004). *Causes of deforestation of the Brazilian Amazon* (Vol. 22). World Bank Publications.

Methodhome. (2017). *Green Glossary*. Retrieved July 18, 2017, from https://www.methodhome.com.au/beyond-bottle/green-glossary/

Mitchell, R. B. (2010). *International politics and the environment*. Los Angeles, CA: Sage Publications.

Najam, A. (2005). Developing countries and global environmental governance: From contestation to participation to engagement. *International Environmental Agreement: Politics, Law and Economics*, 5(3), 303–321. doi:10.100710784-005-3807-6

Namastesolar. (2017). *Operations & Maintanance*. Retrieved from http://www.namastesolar.com/

Newbelgium. (2017). *New Belgium Brewing Company*. Retrieved July 18, 2017, from http://www.newbelgium.com/

Nqa. (2017). *EMAS: EU Eco-Management and Audit Scheme*. Retrieved July 19, 2017 from https://www.nqa.com/en-gb/certification/standards/emas

Pataki, G. E., & Crotty, E. M. (2017). *Understanding and Implementing an Environmental Management System: A step by step guide*. New York State Department of Environmental Conversation Pollution Prevention Unit. Retrieved July 18, 2017, from http://www.dec.ny.gov/docs/permits_ej_operations_pdf/p2emsstep1.pdf

Petheram, L. (2002). *Our Planet in Peril: Acid rain*. Bridgestone Books.

Pickering, K. T., & Owen, L. A. (2006). *An introduction to global environmental issues* (2nd ed.). London: Routledge.

Popescu, F., & Lonel, I. (2010). Anthropogenic air pollution sources. In *Air quality*. InTech. doi:10.5772/9751

Purdey, S. J. (2010). *Economic growth, the environment and international relations: the growth paradigm*. London: Routledge.

Qin, B., Liu, Z., & Havens, K. (Eds.). (2007). *Eutrophication of shallow lakes with special reference to Lake Taihu, China* (Vol. 194). London: Springer. doi:10.1007/978-1-4020-6158-5

Seventhgeneration. (2017). *Nurture Nature*. Retrieved July 18, 2017, from https:// www. seventh generation. com/nurture-nature

Sheldon, C., & Yoxon, M. (2006). *Environmental management systems: a step-by-step guide to implementation and maintenance* (3rd ed.). Routledge.

Shortle, J. S., & Abler, D. G. (Eds.). (2001). *Environmental policies for agricultural pollution control*. New York: CABI Publishing. doi:10.1079/9780851993997.0000

Smith, G. L., Sundaram, S. K., & Spearing, D. R. (2002). *Environmental Issues and Waste Management Technologies in the Ceramic and Nuclear Industries VII, Ceramic Transactions* (Vol. 132). The American Ceramic Society.

Solberg. (2017). *Serious about Sustainability*. Retrieved July 18, 2017, from http:// www. solbergmfg.com/

Somlyódy, L., & van Straten, G. (Eds.). (2012). *Modeling and managing shallow lake eutrophication: with application to Lake Balaton*. London: Springer.

Sulphey, M. M., & Safeer, M. M. (2015). *Introduction to Environment Management* (3rd ed.). Delhi: PHI Learning Pvt. Ltd.

Sungevity. (2017). *Solar Basics*. Retrieved from http: //www. sungevity.com/home/solar-panel-savings?_ga= 2.224662425.285111617.1500500164-1897408410.1500394429

Sunpower. (2017). *Positive Energy Solar*. Retrieved from https:// www. positiveenergysolar. com/

Tinsley, S. (2002). EMS models for business strategy development. *Business Strategy and the Environment*, *11*(6), 376–390. doi:10.1002/bse.340

Tiwary, A., & Colls, J. (2010). *Air pollution: Measurement, modeling, and mitigation* (3rd ed.). London: Taylor & Francis.

UNEP. (2017). *UNEP / FIDIC / ICLEI Urban Environmental Management: Environmental Management Training Resources Kit*. Retrieved July 18, 2017, from http: //www. unep.or.jp /ietc/Announcements /EMSkit_launch.asp

Vajpeyi, D. K. (2001). *Deforestation, environment, and sustainable development: a comparative analysis*. Greenwood Publishing Group.

WasteZero. (2017). *The Trash Problem*. July 18, 2017, from http:// wastezero. com/

Weiß, P., & Bentlage, J. (2006). Environmental Management Systems and Certification. Baltic University Press.

WestPaw. (2017). *Eco-friendly materials*. Retrieved from https: //www. westpaw. com/

Yakhou, M., & Dorweiler, V. P. (2004). Environmental accounting: An essential component of business strategy. *Business Strategy and the Environment*, *13*(2), 65–77. doi:10.1002/bse.395

Chapter 5
Consumer Cooperation in Sustainability:
The Green Gap in an Emerging Market

Njabulo Mkhize
University of kwaZulu-Natal, South Africa

Debbie Ellis
University of KwaZulu-Natal, South Africa

ABSTRACT

The planet is under threat. Unless all stakeholders, that is governments, businesses, and consumers, become more environmentally friendly, some predict dire consequences for the earth and all those who inhabit it. While governments and businesses have a role to play, green consumer behavior is vital to the sustainability of the environment. Consumers have been shown to express increasing concern for the environment, but in many studies this concern has been found not to be matched by actions, a phenomenon labelled the green gap. This chapter describes a study that investigated the existence and extent of the green gap amongst a sample of South African adult consumers and sought to determine possible reasons for a lack of green behavior. Recommendations are made to marketers and policy makers to encourage consumer cooperation in environmental sustainability.

INTRODUCTION

António Guterres, the United Nations Secretary General, warns that climate change and competition for scarce natural resources threaten peace and sustainable development globally and that protecting the environment and having an emphasis

DOI: 10.4018/978-1-5225-3990-2.ch005

on environment prevention methods is imperative (United Nations, 2016). There are two main drivers of the human impact on the environment, namely population and consumption (de Sherbinin, Carr, Cassels, & Jiang, 2007). The last century has seen an increase in the global population, which has resulted in an increase in the food consumption and a deterioration of the environment caused by over consumption and an increase in the use of natural resources (Chen & Chai, 2010). The global population has more than tripled in the last century from an estimated 1.5 living on earth in 2003 to 6.3 billion people and there is a prediction that the population will reach 9.7 billion by 2050 (United Nations, 2015). Most of this growth will come from emerging markets.

From a marketing perceptive an increase in the number of people and their consumption levels may be seen as advantageous since products are being consumed. However thirty to forty percent of the environmental issues are caused by the consumption patterns of human beings (Chan, 2001) thus the decisions and behaviours of consumers have a significant impact on the environment (Tobler, Visschers, & Siegrist, 2012). Climate change, global warming, deforestation, pollution and depletion of the ozone layer are just some of the global environmental crises facing the world (Juwaheer, 2005). These environmental issues give rise to environmental concern amongst governments, activist groups, some businesses and some consumers, which has led to the emergence of the green economy consisting of green or pro-environment governments, companies, marketers and consumers. Global consumers' cooperation in achieving global sustainability is arguably more important than government or business actions as consumers have the greatest impact on sustainability or the lack thereof.

In most studies on environmental concern, it has been found that people are concerned with their impact on the environment and the deterioration thereof (Takács-Sánta, 2007) and these levels of concern have been increasing especially in the last three decades (Kim & Choi, 2005; Synodinos & Bevan-Dye, 2014). This increase in environment concern should translate into green consumer behaviour, i.e. the purchase and use of products and services which are environmental friendly. While some prior research studies indicate a significant positive relationship between environmental concern and green consumer behaviours (Davari & Strutton, 2014; Kim & Choi, 2005) with some even finding environmental concern to be a predictor of green consumer behaviours (Sinnappan & Rahman, 2011), this is not always the case. Other researchers have found that more consumers in the modern society are concerned about the environment than has translated into environmentally friendly behaviours and the purchase of green products (Fisher, Bachman, & Bashyal, 2012; Flynn, Bellaby, & Ricci, 2009; Young, Hwang, McDonald, & Oates, 2010). This gap between green concern and green behaviour is known as the 'green gap' (Kennedy, Beckley, McFarlane, & Nadeau, 2009; Nielsen, 2011).

The efforts made by green consumers to purchase green products and support green initiatives have been hindered by several constraints to adopting green lifestyles. Addressing these constraints will assist in decreasing the green gap globally. This research investigates the extent of the green gap in an emerging market and surfaces some of the perceptions respondents have regarding green behaviour which might indicate reasons these consumers give for not adopting green lifestyles. de Barcellos, Krystallis, de Melo Saab, Kügler, and Grunert (2011) indicate that research on the green gap has been scarce in developing countries, such as South Africa. Without the cooperation of emerging market consumers in achieving sustainability, the efforts of developed nations' governments, business and consumers is unlikely to be enough to save the planet. It will ultimately be vital to garner the support and cooperation of the emerging market consumers as they represent 80% of the world's population (European Central Bank, 2017).

The mission of this chapter is to expose the reader to research conducted on emerging market consumers to determine the existence of, and reasons given for, the green gap among these consumers. After reading this chapter, the reader will understand a key sustainability concern which is understanding of the challenges to getting emerging market consumer cooperation in achieving global sustainability goals. The reader will also be exposed to recommended solutions to enhance green behaviour among these consumers so as to ultimately better achieve global sustainability.

THEORETICAL FRAMEWORK

Various social-psychological theories have been used by researchers of green behaviour to understand this behaviour (Stern, 2000). However, the Knowledge Attitude-Behaviour model has been used widely to describe behaviour change (Derzon & Lipsey, 2002). In the context of green behaviour it represents a linear model which explains that environmental awareness (knowledge) influences environmental concern (attitude) which in turn influences pro-environmental behaviour (Kamate et al., 2009). This study focuses specifically on the concern-behaviour link in the model by investigating the existence of a gap between concern and behaviour and possible reasons for this gap.

Green Consumer Behaviour

There are several terms which have been used in past studies to describe green consumer behaviour such as 'ecological conscious consumer behaviour (ECCB)' (Straughan & Roberts, 1999), 'pro-environmental consumer behaviour' (Kaufmann, Panni, &

Orphanidou, 2012), 'environmentally conscious consumer behaviour'(Gan, Wee, Ozanne, & Kao, 2008), 'environmentally friendly consumer behaviour'(Thøgersen & Ölander, 2003) and 'green consumer behaviour'(Jansson, Marell, & Nordlund, 2010). This section uses these names interchangeably.

Green consumers are understood to be people who have a particular interest in, and awareness of 'green' issues and make an ongoing attempt to change their behaviours to protect the environment (Chen & Chai, 2010). Moisander (2007) states that green consumers are characterised by the willingness to use their purchasing power to cause a social change. With the purpose of reducing the negative effect on the environment they engage in behaviours of recycling materials, buying green products, being aware of and supporting companies which are green, checking labels and packaging to see if they are environmentally friendly and adopting behaviours which lessen or reduce green issues (Albayrak, Caber, Moutinho, & Herstein, 2011; Laroche, Bergeron, & Barbaro-Forleo, 2001) such as conserving energy and water and supporting conservation campaigns such as 'Save the Rhino'.

Environmental Concern

Environmental concern or concern for the natural environment (Hartmann & Apaolaza-Ibáñez, 2012) can be defined "as a global attitude with indirect effects on behaviour through behavioural intention" (Kaufmann et al., 2012, p. 53). Environmental concern is an important attitude in the struggle for the sustainability of the environment. According to Pagiaslis and Krontalis (2014), ecological or environmental concern is a concept of general belief that is an antecedent of more specific concepts such as green knowledge and environmental beliefs associated with green products. The levels of environmental concern across the globe have been increasing steadily. In America for example, studies show that there was an increase in environmental concern from 62% to 77% between 2004 and 2006 (Ryan, 2006) and in a recent study by van Riper and Kyle (2014) high levels of environmental concern amongst US respondents were found. Environmental concern has also become important for the international community too. In a study in Hong Kong, 80% of respondents expressed dissatisfaction with the city's environmental quality (Lee, 2008) indicating concern for the environment. In the United Kingdom, approximately 47% of consumers have favourable attitudes towards organic foods (Albayrak et al., 2011). Studies of various aspects of environmental concern in several populations in South Africa have found environmental concern for pollution, waste (Rousseau & Venter, 2001), air and water (Berndt & Petzer, 2011; Davari & Strutton, 2014; Kim & Choi, 2005; Straughan & Roberts, 1999) however levels of environmental concern have generally been found to be lower in developing nations than in developed nations (Paul, Modi & Patel, 2016).

In some cases a positive correlation has been found between the environmental concern and environmentally friendly behavior, and environmental concern has been found to predict green behavior (Kaufmann et al., 2012; Lee, 2008; Sinnappan & Rahman, 2011). However this concern does not always match behaviour and a green gap has been found between concern and behaviour.

The Green Gap

Generally individuals consider themselves as eco-friendly or sensitive to the environment but their behaviour often contradicts their confessed sensitiveness (Albayrak et al., 2011; Fisher et al., 2012). This gap has been called the value-action gap (Honabarger, 2011), the environmental values-behaviour (EVB) gap (Kennedy et al., 2009), and the 'green gap' (Honabarger, 2011).The 'green gap' can also be defined as "the observed disparity between people's reported concerns about key environmental, social, economic or ethical concerns and the lifestyle or purchasing decisions that they make in practice" (Flynn et al., 2009, p. 158).

Evidence of the green gap: There is compelling evidence supporting the existence of the green gap at least in developed nations. For example, in a Canadian study done in 2004, it was shown that 72.3% of the respondents stated they were concerned about the environment but their actions did not support their concern as only 10.8% stated their lifestyles were supportive of the environment (Kennedy et al., 2009, p. 156). Similarly, a global survey done by McKinsey & Company indicated that 87% of the participants were concerned about the environmental and social impact caused by their purchase behaviour but only 33% admitted to have bought products that assist in sustaining the environment (Bonini & Oppenheim, 2008). More recently in a study by Nielsen, it was shown that 83% of consumers in the world are aware of the need to implement programmes to assist with environmental recovery yet less than 20% are willing to pay more for their green products (Nielsen, 2011).

The above studies mostly reflect research done in developed countries. Few studies have been conducted in developing nations. A study conducted in Turkey which looked at consumer skepticism towards green products found a disparity between green consumer behaviour and environmental concern (Albayrak et al., 2011). Another study conducted in India explored the green concern and behaviour in the lodging industry and found that respondents were conscious of environmental issues but they still did not adopt green lifestyles, because they were not willing to pay a premium price (Manaktola & Jauhari, 2007). Pantelic, Sakal, and Zehetner (2016) found that their Serbian sample also exhibited positive attitudes towards sustainability but lacked the corresponding actions. In South Africa, some studies have been done on environmental perceptions, awareness and behaviour (e.g. Anderson, Romani, Phillips, Wentzel, & Tlabela, 2007). A study conducted in the

Durban Municipality researched making laws at local government level to adapt to climate change (Roberts, 2008). However, no studies could be found that specifically investigate the green gap in South Africa. This study aims to fill this research gap.

Reasons for the Gap: Understanding green behaviour is far more complicated than previously thought (Gifford & Nilsson, 2014) and determining the reasons or the barriers for consumers to adopt green lifestyles has been said to be a complex issue (Kaufmann et al., 2012). While individuals can state that they value the environment, in some instances other issues take precedence over their concern for the environment e.g. financial security and safety (Kennedy et al., 2009). Han, Hsu, and Lee (2009) state that monetary and non-monetary costs are reasons for the small numbers of eco-friendly consumers who are in the market. Several reasons have been proposed for why consumers do not adopt ecologically conscious consumer behaviour:

- **Green Products Are Perceived to be Expensive:** Green products are, and are perceived to be, more expensive yet provide the same functions (Bennet & Williams, 2011). Although long-term cost savings may come from using energy efficient products like hybrid cars and light bulbs, initial costs are usually higher (Stephenson & Rodriquez, 2014). Majláth (2010) states that price is the biggest reason that consumers do not buy green products. A study conducted in India found high prices as a major barrier to organic food consumption as Indian customers are price sensitive compared to developed nations (Yadav & Pathak, 2016, p. 123).
- **Green Product Availability:** Takács-Sánta (2007) argues that it is not the weakness of environmental concern but the lack of alternatives, which jeopardises adoption of ecologically conscious behavior. Tilikidou (2007) found that some green products, such as organic wine, toiletries, pasta and clothing, are difficult to find. GrailResearch (2011) reports that 34% of their US respondents indicated low availability of green products as a reason for not buying these products. In India, also a developing market, green choices were perceived to be low (Paul et al., 2016).
- **Perception of Inferiority of Green Products in Terms of Quality and Efficiency:** Sharma (2011) states that one of the biggest obstacles to consumers' adopting a green lifestyle or purchasing eco-friendly products is the lack of trust in green products. Consumers' scepticism, or distrust of green products, reduces their ability to make a pro-environmental decision (Albayrak et al., 2011).
- **Perceived Consumer Effectiveness:** Perceived Consumer Effectiveness (PCE) is defined as the extent to which consumers believe that their individual efforts can have a positive impact on the environment (Chang, 2011; Kim & Choi, 2005). PCE is seen to be a significant predictor of a variety of

ecologically conscious and pro-environmental consumer behaviours (Chang, 2011; Majláth, 2010; Rex & Baumann, 2007; Straughan & Roberts, 1999; Vicente-Molina, Fernández-Sáinz, & Izagirre-Olaizola, 2013). Roberts (1996) argues that PCE is the single strongest predictor of green consumer behaviour, surpassing both psychological and demographic factors.

- **Lack of Information About the Availability of Green Products:** Lack of information is one factor affecting the practicality of adopting a green lifestyle (Kollmuss & Agyeman, 2002).

These reasons are investigated in this study of South African consumers. Soyez (2012) states that in terms of environmental research, researchers ought to concentrate more on examining emerging economies that have an effect on the environment such as South Africa in order to understand the actions of consumers better.

A STUDY OF EMERGING MARKET CONSUMERS

Research Methods

A descriptive research design using quantitative data was used to describe participants concern for the environment and green behaviour so as to determine the existence and extent of the green gap, as well as the reasons for a lack of green behaviour. The questionnaire was developed to measure the constructs of this study. Ethical clearance was obtained from the university's Research Office before the commencement of the study. All respondents were assured of their anonymity and confidentiality.

Measures

The questionnaire included the environmental behaviour questions from the ECCB scale developed by Roberts (1996) and used by Straughan and Roberts (1999) and more recently by several authors (e.g. Datta, 2011; McEachern & Carrigan, 2012). To measure environmental concern the Environmental Concern scale, also called the New Ecological Paradigm (NEP) scale developed by Dunlap, Kent, Angela, and Robert (2000) was used. This scale has also been used in many other studies (e.g. Amburgey & Thoman, 2012; Vikan, Camino, Biaggio, & Nordvik, 2007). The Perceived Consumer Effectiveness scale items were taken from the Straughan and Roberts (1999) questionnaire. The other questions on reasons for the lack of ECCB, were developed from the literature review. Previous use of the main measures indicated that these scales exhibit internal consistency and can thus be deemed reliable. For example, the Cronbach's alpha for the ECCB scale was 0.96, Environmental

Concern scale 0.84 and the Perceived Consumer Effectiveness had an alpha scale of 0.72 (Roberts, 1996, p.223-224). Hair, Bush, and Ortinau (2009) state that a coefficient alpha of less than 0.6 indicates a low to marginal (unsatisfactory) internal consistency. Therefore, looking at the Table 1, the constructs exhibited acceptable internal consistency reliability in this study.

Pilot studies are used to identify weaknesses or limitations of the questionnaire (Hair et al., 2009; Muijs, 2010). The pilot study sample was chosen to represent the research sample. A sample of 30 was selected to represent all the demographic variables relevant to the study in relation to race, gender and age. The sample was conveniently selected and this was seen as reasonable since the study was also using convenience sampling. The pilot study ensured that different types of respondents would be able to understand the research instrument. The pilot study was conducted at the same study location as the main survey. The length of the questionnaire was particularly taken into consideration while conducting the pilot test as it was seen as a critical factor to receiving unbiased data. The longest time it took a respondent to complete the questionnaire was 10 minutes but some respondents were able to complete it in 5 minutes. It was decided that this was an acceptable time for respondents to complete the questionnaire without negatively affecting response rate. Therefore, there were no changes in terms of the length of the questionnaire. Pilot study respondents were also asked to give feedback on the questionnaire. They experienced no difficulty in answering the questions therefore no changes were made to the research instrument. The original Likert scale format was used where 1=*strongly agree* and 5= strongly *disagree*.

The Sample

Since the research investigated the 'green gap', which is the difference between the levels of concern and the purchasing behaviour of consumers, an adult population was seen as an appropriate population since they possess purchasing power. In the South African constitution any person older than 18 is considered to be an adult, and thus has the capacity to make their own decisions (Strode, Slack, & Essack, 2010). The study took place in a municipality in South Africa with approximately

Table 1. Cronbach alpha scores for the current research instrument

Scale	Cronbach Alpha
Ecologically conscious consumer behavior (ECCB)	0.88
Perceived Consumer Effectiveness (PCE)	0.67
Environmental Concern (EC)	0.76

423 000 people who are over the age of 15 with 53% female and 47% male. The race distribution of residents is approximately 81% African, 10% Indians, 6% White and 3% Coloured (StatisticsSA, 2011).

Since the questionnaire included a large number of constructs each with multiple measures, it was fairly lengthy, thus, needed to be distributed to respondents where it would be convenient for them to complete. The study site chosen was the Mkondeni Test Driving Centre (MTDC) Licensing Department. This study site was chosen because it is the place where all residents have to go to book their learners and driving tests, to be tested for their learner and driving licenses, and to renew their licenses every 5 years, and thus, it provided access to a large number of relevant members of the target population. There are also usually fairly long queues thus providing people who may be more likely to have time to answer the questionnaire, thereby improving the response rate. The MTDC is ideal for data collection as it is the only testing centre in the city and the people who go there are over the age of 18 years, due to legal age restriction for driving. It is acknowledged that any study site is likely to have some limitations. In this case the site may exclude certain elements of the population, for example, those who do not drive. Overall it was believed that the benefits of the site outweighed the limitations, as the site provided a good cross-section of respondents from all race groups, genders and ages above 18. Other study sites such as shopping malls were considered to be inferior as they have been over-used in consumer behaviour studies in the past and are also unsuitable for lengthy questionnaires. Data was collected over a period of a month, 3 days per week, chosen to ensure all days were eventually covered during the month.

At the confidence level of 95% and with a 5% confidence interval, a sample of 378 was desired. Using non-probability, convenience sampling 378 questionnaires were distributed and 330 questionnaires returned of which 317 were usable. This gave an effective response rate of 84%. The demographic profile of the sample is presented in Table 2.

As illustrated in Table 2, the gender and race profile is not too dissimilar to that of the municipality's population, although the sample is somewhat skewed in terms of age with the majority of respondents being in the age groups 18 to 30. In terms of the level of education of this sample, it indicated a highly literate group as the majority (98.4%) of respondents hold a National Senior Certificate (NSC) qualification and above. The NSC qualification is the national school leaving certificate after 12 years of schooling. This is in contrast to the municipality statistic indicating that only approximately 24% of the municipality's population hold a NSC or higher (StatisticsSA, 2011). It should be kept in mind however that the study was restricted to adults while the city statistics reflect all ages. Thus, overall it is felt that this sample bears close resemblance on demographics, to the city's adult population.

Table 2. Demographic profile of the sample

Demographics	Frequency (Percentage)						Missing
Gender	Male	Female					
	160(50.5)	144(45.4)					13 (4.1)
Education	Below matric	NCS	Diploma/degree	Postgrad			
	5 (1.6)	122(38.5)	144(36)	60(18.9)			16(5)
Place of residence	Rural	Township	CBD	Suburban			
	41(12.9)	61(19.2)	41(12.9)	182(47.9)			22 (6.9)
Race	African	White	Indian	Coloured	others		
	233 (73.5)	16 (5)	37 (11.7)	13(4.1)	2(0.6)		16(5)
Age	18-20	21-30	31-40	41-50	50 &above		
	109 (34.4)	160 (50.5)	12 (3.8)	12 (3.8)	6 (1.9)		18(5.7)
Household income	Under 40k	40001-80k	80001-160k	160001-320k	320001-640k	1.2million & above	
	99(31.2)	39(12.3)	46(14.5)	21(8.5)	46(14.5)	8 (2.5)	52 (16.5)

Results

After reversing negatively worded item scores, composite scores were created for each construct representing the mean score for the construct i.e. for Environmental Concern (EC), Ecologically conscious consumer behaviour (ECCB) and Perceived Consumer Effectiveness (PCE).

The Green Gap

As with other studies on the green gap (e.g. Nielsen, 2011), the difference between ecologically conscious consumer behaviour (ECCB) and environmental concern (EC), was calculated. In terms of an equation, the Green Gap = ECCB - EC (Nielsen, 2011). If the mean of the ECCB construct is greater (i.e. showing more green behaviour) than the mean for the EC construct, then a positive gap exists (there is no 'green gap') and there is more action towards a sustainable environment than there is concern. The green gap occurs when the ECCB mean is less than the EC mean. The results of this study indicate a green gap of -.661. Thus concern is greater than behaviour. This difference in composite scores is significant t (58) = -6.671, p < .000(2-tailed) and the eta squared statistic (.43) indicates a large effect size. This suggests that the respondents of this study have higher levels of concern than behaviour. This is consistent with a global study conducted by Nielsen which found that 87% of the participants were found to be concerned about the environment but only 33% converted their concern into action when they bought products. In the current study 73% of respondents were concerned but only 22% converted their environmental concern into green consumer behaviour. Possible reasons for the lower levels of behaviour were then explored.

Reasons for the Green Gap

This section is divided into two parts, firstly a discussion of Perceived Consumer Effectiveness as a possible reason for lack of ECCB, and secondly other reasons for the lack of ECCB.

Perceived Consumer Effectiveness (PCE) is defined as the extent to which consumers believe that their individual efforts can make a positive impact on the environment (Chang, 2011; Kim & Choi, 2005). Thus consumers may not behave in an environmentally-friendly manner because they do not believe their actions will make a difference (Straughan & Roberts, 1999). In this study the composite mean score for PCE was 2.17 (STD = .691) where 1 represents a highest PCE and 5 the lowest PCE. Fifty eight percent of the respondents, as well as the relatively low mean (2.17) indicates that respondents believe that their individual actions can make a difference towards the sustaining of the environment. Thus it seems that a lack of perceived effectiveness should not be the reason for the low levels of ECCB. Other reasons are investigated below.

Table 3 indicates that the reasons for the lack of green consumer behaviour in this sample were predominantly the perceptions of a lack of promotion of green

Table 3. Other reasons for the green gap

Reasons for the 'Green Gap' Statement	ATSE	N	DTSE	Missing	Mean	Std dev.
Product quality of green products is inferior	84 (26.5)	98 (30.9)	125 (39.5)	10 (3.1)	3.22	1.21
Promotion of eco-friendly products is lacking	197 (62.1)	56 (17.7)	58 (18.3)	6 (1.9)	2.34	1.09
Unable to distinguish between green and conventional products	104 (32.8)	94 (29.7)	109 (34.4)	10 (3.1)	3.03	1.14
Green products are not easily available	152 (38)	81 (25.6)	78 (24.8)	6 (1.9)	2.66	1.09
I am not aware of any green products	64 (20.2)	60 (18.9)	188 (59.3)	4 (1.3)	3.55	1.14
I don't trust green branded products	45 (14.2)	77 (24.3)	191 (60.2)	4 (1.3)	3.65	1.01
Green products are clearly labelled	102 (32.1)	88 (27.8)	121 (58.2)	6 (1.9)	3.14	1.11
There are no or limited mechanisms available to practice a green lifestyle	130 (41)	99 (31.2)	82 (25.9)	6 (1.9)	2.77	1.11

ATSE= Agree to some extent; N= neutral; DTSE= Disagree to some extent products (M=2.34), the lack of availability of green products (M=2.66), that green products are not clearly labelled (M=3.14) and the lack of mechanisms to support a green lifestyle (2.77) where 1 = strongly agree and 5 = strongly disagree.

BUSINESS AND POLICY IMPLCATIONS AND RECOMMENDATIONS

Lee (2008) argues that environmental concern is the second top predictor of green purchasing behaviour. In an application to the eco-hospitality industry, Han et al. (2009) found that there is a significant relationship between level of environmental concern and the number of people who intend to stay at an Eco-hotel. Similarly Laroche et al. (2001) state that there is positive relationship between the willingness of consumers to pay a premium price and the extent of environmental concern. Thus it seems environmental concern may be a necessary prerequisite to environmental behaviour. Similar to other studies (e.g. Lee, 2008; Nielsen, 2011) the current study found that respondents showed high levels of concern about the environment. However it is behaving in an environmentally friendly way that is the required consumer cooperation essential in the fight against the depletion of the environment. Concern may be necessary but studies have found that it is not always sufficient to ensure behaviour and a green gap has often been found between concern and behaviour (Laroche et al., 2001; Nielsen, 2011). In this study, ECCB item scores indicate that respondents have participated in some actions that are pro-environmental e.g. reducing electricity consumption but on the whole, green behaviour lags behind concern. Only 13% of respondents can be said to frequently act in a green manner. Although very low, this figure is not totally dissimilar to a similar global study where the percentage was 22% (Nielsen, 2011). Thus like many other countries, this study of South African consumers also indicates a significant gap between environmental concern and behaviour.

The second objective of this research was to try to understand why green behaviour may be lacking in developing nations such as South Africa. Barriers to adopting of the green lifestyle are evident in prior studies (e.g. Kollmuss & Agyeman, 2002; Nielsen, 2011). The participants of this study also indicated that these barriers impede their green behaviour. The main reason for not behaving in a green manner appears to be a perception of a lack of promotion of green products. If consumers are not made aware of the existence of green products or of the benefits of these products to consumers and the environment, then consumers cannot be expected to cooperate in achieving global sustainability by purchasing green products. Promotion is a fundamental aspect of the marketing mix and necessary for the adoption of these products. Better information by organisations that are promoting the adoption of green products and services is vital to gaining the trust of consumers in the market (Ozaki, 2011). Producers of green products need to address this problem by actively promoting their products so that consumers recognise green products as strong alternatives to traditional products which are more harmful to the environment. Money is needed to aggressively promote green products in all forms of media. An integrated marketing communication plan needs to be designed and implemented.

Eco-labels are also a good way to notify consumers that products are environmentally friendly. According to Rashid (2009, p.134) eco-labels are eye-catching tools that notify consumers. The Global Ecolabelling Network states that an eco-label is a label which "identifies overall environmental preference of a product or service within a specific product/service category based on life cycle considerations" (Global Ecolabelling Network, 2011, para4). To be of use in steering consumers towards green products, these labels have to be credible, meaningful and recognisable to consumers (D'Souza, Lamb & Taghian, 2006). Third party environmental labelling has superior reliability as they involve a 3rd party evaluation of a company's environmental principles (D'Souza et al., 2006). Once a manufacturer passes the certification process the product is awarded a seal or symbol which is a sign that the producer's product has been deemed environmentally safe (Engels, Hansmann & Scholza, 2010). As independent verification is used to determine if a manufacturer meets the criteria, the consumers trust these labels more in comparison to the manufacturers' self-installed labels (Grankvist, Dahlstrand & Biel, 2004). Consumers' assessment of environmental performance of brands is usually affected completely by environmental symbols (Roe, Teisl & Rong, 2001). Unfortunately there is no independent third party to officially certify environmentally friendly products in South Africa (Scott & Vigar-Ellis, 2014) however these authors recommend organisations apply to use international symbols or alternatively use terms such as non-harmful, biodegradable or recyclable on packaging as these were found in their South African study, to be associated with environmentally friendly packaging.

The government also has a role to play here. Firstly, the development of a national environmental certification process and symbol would ensure that environmental claims made by companies are legitimate. Secondly, having one symbol that consumers can look out for and trust will assist them in identifying truly green products. This is important as Brennan and Binney (2008) state that green washing (the false claims made by companies about the greenness of their products) has resulted in consumer skepticism. Scott and Vigar-Ellis (2014) also state that where there is no national symbol and certification is industry specific, consumers are likely to become confused by what constitutes environmentally friendly products. Thus to alleviate green washing, consumer skepticism and consumer confusion, and national certification system and symbol is required.

Consumer Cooperation in Sustainability

A similar potential reason for the lack of green behaviour is the perceived lack of availability of green products in the marketplace. While the above discussion assumes green products exist but are not effectively promoted, here the respondents indicate a perception that there are not many green products available. While the number of green alternatives is likely to be less than in developed countries, green products certainly do exist and the challenge to the producers of these products is to increase consumer awareness of these and to make them more readily available to consumers. This may include more aggressive advertising of these products as mentioned above but distribution of green products also plays an important role in giving a competitive advantage (Smith & Perks, 2010). Improved availability of green products means that consumers need to find green products whenever they need them. Green products have to be in close proximity to consumers. This can be done by each store having a green product shelf easily visible in the supermarket. Green product manufacturers can also investigate joint awareness-building campaigns undertaken with leading retailers or distributers of these products and more intensive distribution. Certain retailers, such as Woolworths have positioned themselves as supporting the green initiative and thus green products assisting them to achieve this position while maintaining product quality standards may be well received. Of course organisations that have not produced green alternatives can look at greening their processes and products to make more green products available.

A major aspect of green behaviour is green activities as opposed to the purchasing of green products which only represents one aspect of green behaviour. Green activities are part of a broader green lifestyle and include reducing the use of non-renewable or scarce resources e.g. electricity, fuel and water; recycling and reusing; and supporting animal and plant conservation campaigns. The second most likely reason for a lack of green behaviour amongst this sample of South African consumers was the perceived lack of mechanisms to practice a green lifestyle. A lack of the necessary infrastructure or a civic movement promoting and supporting green behaviour may well cause the green gap (Pantelic et al., 2016). While businesses may support these initiatives, these mechanisms often need to be supplied, or initiated by government or non-profit organisations. Thus all parties in society need to share the responsibility of conserving the planet. However, specifically in terms of mechanisms to support a green lifestyle, a basic example of such mechanisms is the provision of recycling stations. In comparison to most developed countries, these are sorely lacking in South Africa. Social marketing campaigns can also be a critical marketing activity required here. Eskom, South Africa's only supplier of electricity, ran a media campaign aimed at convincing the general public to reduce their electricity consumption during the peak demand hours of 18h00 to 21h00. The Power Alert campaign was successful in reducing peak time load between 100 and 900 megawatts, depending on the level of criticality (Etzinger, 2011). Eskom's provision of energy efficient light bulbs to replace traditional bulbs was also successful. This initiative has seen the rollout of 54 million compact fluorescent lamps saved more than 1000 megawatts (Eskom, 2015). Thus more needs to be done by government at national and local level to make it an easy option for consumers to adopt the green lifestyle.

LIMITATIONS AND DIRECTIONS FOR FUTURE RESEARCH

As with most study sites, the chosen site had some limitations in that the sample appeared to be skewed towards a slightly younger and more educated sample than perhaps other sites would have. Older and less educated respondents may have had different concerns, behaviours and perceptions. The highly literate sample however did have the advantage of minimising language problems with using an English questionnaire with a predominantly African sample for whom English would have been a second language. The NCS qualification which the majority of respondents had attained includes English so translation problems were not believed to hinder the quality of the data. As the combined effort of all consumers is required to achieve global sustainability, future studies should investigate the perceptions, concerns and behaviours of the less literate emerging market consumers. Although this is acknowledged to be a more difficult task, qualitative studies conducted in mother tongues could provide valuable insights into this important, and relatively large, component of emerging markets. The study was focused in one city, in one, relatively small, emerging market and it is possible that other regions may have yielded different results. With the number of studies beginning to be done in emerging markets such as China and India, a comparative study would be useful. Finally, neutral or indecisive scores were recorded for a few of the concern statements. This may need more in-depth investigation. Why were respondents indecisive? Was it because they didn't understand the statement or because there were mediating factors? Environmental concern and behaviours are complex constructs which may be productively investigated using more in-depth qualitative methods. Could for example, relatively higher concerns for poverty, health and safety make green behaviour impractical for emerging market consumers?

CONCLUSION

As in many other countries a green gap exists between the environmental concern and behaviours of this sample of South Africa consumers. Consumer cooperation is arguably more important in sustaining the global environment than the actions of businesses and governments, thus understanding the reasons for this lack of green behaviour is vital in achieving consumer cooperation. In this chapter, a study of emerging market consumers green gap and possible reasons therefore was undertaken. Barriers or inhibitors of environmental behaviour in this emerging market were found to be the perceived lack of green products and promotion of those that do exist, as well as a lack of mechanisms to assist in the adoption of green activities such as recycling, reuse and support of conservation initiatives. These emerging market consumers are concerned about the environment but these barriers prevent the behaviours required to sustain the planet. To assist consumers to do their bit for global sustainability, businesses and governments must play their respective parts as indicated in the recommendations of this chapter. Ultimately a great deal of cooperation between the major stakeholders: governments, businesses and consumers is necessary to have a positive impact on the sustainability of the environment.

This research received no specific grant from any funding agency in the public, commercial, or not-for-profit sectors. Permission for the study was granted by the KZN Road Traffic Inspectorate.

REFERENCES

Albayrak, T., Caber, M., Moutinho, L., & Herstein, R. (2011). The influence of skepticism on green purchase behavior. *International Journal of Business and Social Science*, *2*(13), 189–197.

Amburgey, J. W., & Thoman, D. B. (2012). Dimensionality of the New Ecological Paradigm Issues of Factor Structure and Measurement. *Environment and Behavior*, *44*(2), 235–256. doi:10.1177/0013916511402064

Anderson, B. A., Romani, J. H., Phillips, H., Wentzel, M., & Tlabela, K. (2007). Exploring environmental perceptions, behaviors and awareness: Water and water pollution in South Africa. *Population and Environment*, *28*(3), 133–161. doi:10.100711111-007-0038-5

Bennet, G., & Williams, F. (2011). *Mainstream Green: Moving sustainability from niche to normal*. Chicago: Ogilvy & Mather.

Berndt, A., & Petzer, D. (2011). Environmental concern of South African cohorts: An exploratory study. *African Journal of Business Management*, *5*(19), 7899–7910.

Bonini, S., & Oppenheim, J. (2008). Cultivating the green consumer. *Stanford Social Innovation Review*, *6*(4), 56–61.

Brennan, L., & Binney, W. (2008). Is it green marketing, greenwash or hogwash? We need to know if we want to change things. In *Proceedings of Partnerships, proof and practice: International nonprofit and social marketing conference*. University of Wollongong.

Chan, R. (2001). Determinants of Chinese consumers' green purchase behavior. *Psychology and Marketing*, *18*(4), 389–413. doi:10.1002/mar.1013

Chang, C. (2011). Feeling ambivalent about going green. *Journal of Advertising*, *40*(4), 19–32. doi:10.2753/JOA0091-3367400402

Chen, T. B., & Chai, L. T. (2010). Attitude towards the environment and green products: Consumers' perspective. *Management Science and Engineering*, *4*(2), 27.

D'Souza, C., Taghian, M., & Lamb, P. (2006). An empirical study on the influence of environmental labels on consumers. *Corporate Communication: An International Journal*, *11*(2), 162–173. doi:10.1108/13563280610661697

Datta, S. K. (2011). Pro-environmental concern influencing green buying: A study on Indian consumers. *International Journal of Business and Management*, *6*(6), 124.

Davari, A., & Strutton, D. (2014). Marketing mix strategies for closing the gap between green consumers' pro-environmental beliefs and behaviors. *Journal of Strategic Marketing*, *22*(7), 563–586. doi:10.1080/0965254X.2014.914059

de Barcellos, M. D., Krystallis, A., de Melo Saab, M. S., Kügler, J. O., & Grunert, K. G. (2011). Investigating the gap between citizens' sustainability attitudes and food purchasing behaviour: Empirical evidence from Brazilian pork consumers. *International Journal of Consumer Studies*, *35*(4), 391–402. doi:10.1111/j.1470-6431.2010.00978.x

De Sherbinin, A., Carr, D., Cassels, S., & Jiang, L. (2007). Population and environment. *Annual Review of Environment and Resources*, *32*(1), 345–373. doi:10.1146/annurev.energy.32.041306.100243 PMID:20011237

Derzon, J. H., & Lipsey, M. W. (2002). *A meta-analysis of the effectiveness of mass communication for changing substance use knowledge, attitudes and behavior. In Mass media and drug prevention: Classic and contemporary theories and research* (pp. 231–258). Mahwah, NJ: Erlbaum.

Dunlap, R., Kent, D., Angela, G., & Robert, E. (2000). Measuring Endorsement of the New Ecological Paradigm: A revised NEP Scale. *The Journal of Social Issues*, *56*(3), 425–442. doi:10.1111/0022-4537.00176

Engels, S. V., Hansmann, R., & Scholz, R. W. (2010). Toward a sustainability label for food products: An analysis of experts' and consumers' acceptance. *Ecology of Food and Nutrition, 49*(1), 30–60. doi:10.1080/03670240903433154 PMID:21883088

Eskom. (2015). *Eskom IDM Programme: Focus on Housing Sector of South Africa.* Retrieved 13 May 2017 from http://www.ieadsm.org/wp/files/Ncayiyana_CapeTown.pdf

Etzinger, A. (2011). *Eskom Intergrated demand Management.* Association of Municipal Electrical Utilities (AMEU) Convention. Retrieved 13 May 2017 from http://www.ameu.co.za/Portals/16/Conventions/Convention%202011/Papers/Eskom%20Integrated%20Demand%20Management%20-%20A%20Etzinger.pdf

European Central Bank. (2017). *Growing Importance of Emerging Markets.* Retrieved 13 May 2017 from https://www.ecb.europa.eu/ecb/tasks/international/emerging/html/index.en.html

Fisher, C., Bachman, B., & Bashyal, S. (2012). Demographic impacts on environmentally friendly purchase behaviors. *Journal of Targeting, Measurement and Analysis for Marketing, 20*(3), 172–184. doi:10.1057/jt.2012.13

Flynn, R., Bellaby, P., & Ricci, M. (2009). The 'value-action gap' in public attitudes towards sustainable energy: The case of hydrogen energy. *The Sociological Review, 57*(2_suppl), 159–180. doi:10.1111/j.1467-954X.2010.01891.x

Gan, C., Wee, H. Y., Ozanne, L., & Kao, T.-H. (2008). Consumers' purchasing behavior towards green products in New Zealand. *Innovative Marketing, 4*(1), 93–102.

Gifford, R., & Nilsson, A. (2014). Personal and social factors that influence pro-environmental concern and behaviour: A review. *International Journal of Psychology, 49*(3), 141–157. PMID:24821503

Global Ecolabelling Network. (2011). *What Is Ecolabelling?* Retrieved 13 May 2017 from http://www.globalecolabelling.net/what_is_ecolabelling

GrailResearch. (2011). *The Green Evolution.* Retrieved 13 May 2017 from http://www.grailresearch.com/pdf/Blog/Grail-Research-Green-Evolution-Study_240.pdf

Grankvist, G., Dahlstrand, U., & Biel, A. (2004). The impact of environmental labelling on consumer preference: Negative vs. positive labels. *Journal of Consumer Policy, 27*(2), 213–230. doi:10.1023/B:COPO.0000028167.54739.94

Hair, J., Bush, R., & Ortinau, D. (2009). *Marketing Research: In a digital information environment* (4th ed.). Boston: McGraw-Hill.

Han, H., Hsu, L.-T. J., & Lee, J.-S. (2009). Empirical investigation of the roles of attitudes toward green behaviors, overall image, gender, and age in hotel customers' eco-friendly decision-making process. *International Journal of Hospitality Management*, *28*(4), 519–528. doi:10.1016/j.ijhm.2009.02.004

Hartmann, P., & Apaolaza-Ibáñez, V. (2012). Consumer attitude and purchase intention toward green energy brands: The roles of psychological benefits and environmental concern. *Journal of Business Research*, *65*(9), 1254–1263. doi:10.1016/j.jbusres.2011.11.001

Honabarger, D. (2011). *Bridging the Gap: The Connection Between Environmental Awareness, Past Environmental Behavior, and Green Purchasing* (Masters thesis). American University, Washington, DC. Retrieved 13 May 2017 from http://www.american.edu/soc/communication/upload/Darcie-Honabarger.pdf

Jansson, J., Marell, A., & Nordlund, A. (2010). Green consumer behavior: Determinants of curtailment and eco-innovation adoption. *Journal of Consumer Marketing*, *27*(4), 358–370. doi:10.1108/07363761011052396

Juwaheer, T. D. (2005). An emerging environmental market in Mauritius: Myth or reality? *World Review of Entrepreneurship, Management and Sustainable Development*, *1*(1), 57–76. doi:10.1504/WREMSD.2005.007753

Kamate, S. K., Agrawal, A., Chaudhary, H., Singh, K., Mishra, P., & Asawa, K. (2009). Public knowledge, attitude and behavioural changes in an Indian population during the Influenza A (H1N1) outbreak. *The Journal of Infection in Developing Countries*, *4*(1), 7-14.

Kaufmann, H. R., Panni, M., & Orphanidou, Y. (2012). Factors affecting consumers' green purchasing behavior: An integrated conceptual framework. *Amfiteatru Economic*, *14*(31), 50–69.

Kennedy, E. H., Beckley, T. M., McFarlane, B. L., & Nadeau, S. (2009). Why we don't" walk the talk": Understanding the environmental values/behaviour gap in Canada. *Human Ecology Review*, *16*(2), 151.

Kim, Y., & Choi, S. M. (2005). Antecedents of green purchase behavior: An examination of collectivism, environmental concern, and PCE. *Advances in Consumer Research. Association for Consumer Research (U. S.), 32*, 592.

Kollmuss, A., & Agyeman, J. (2002). Mind the gap: Why do people act environmentally and what are the barriers to pro-environmental behavior? *Environmental Education Research, 8*(3), 239–260. doi:10.1080/13504620220145401

Laroche, M., Bergeron, J., & Barbaro-Forleo, G. (2001). Targeting consumers who are willing to pay more for environmentally friendly products. *Journal of Consumer Marketing, 18*(6), 503–520. doi:10.1108/EUM0000000006155

Lee, K. (2008). Opportunities for green marketing: Young consumers. *Marketing Intelligence & Planning, 26*(6), 573–586. doi:10.1108/02634500810902839

Majláth, M. (2010). *Can Individuals do anything for the Environment? - The Role of Perceived Consumer Effectiveness.* Paper presented at the FIKUSZ '10 Symposium for Young Researchers, Budapest, Hungary.

Manaktola, K., & Jauhari, V. (2007). Exploring consumer attitude and behaviour towards green practices in the lodging industry in India. *International Journal of Contemporary Hospitality Management, 19*(5), 364–377. doi:10.1108/09596110710757534

McEachern, M. G., & Carrigan, M. (2012). Revisiting contemporary issues in green/ethical marketing: An introduction to the special issue. *Journal of Marketing Management, 28*(3-4), 189–194. doi:10.1080/0267257X.2012.666877

Moisander, J. (2007). Motivational complexity of green consumerism. *International Journal of Consumer Studies, 31*(4), 404–409. doi:10.1111/j.1470-6431.2007.00586.x

Muijs, D. (2010). *Doing quantitative research in education with SPSS.* Thousand Oaks, CA: Sage Publications.

Nielsen. (2011). *The 'Green' gap between environmental concerns and the cash register.* Retrieved 13 May 2017 from http://www.nielsen.com/us/en/insights/news/2011/the-green-gap-between-environmental-concerns-and-the-cash-register.html

Ozaki, R. (2011). Adopting sustainable innovation: What makes consumers sign up to green electricity? *Business Strategy and the Environment, 20*(1), 1–17. doi:10.1002/bse.650

Pagiaslis, A., & Krontalis, A. K. (2014). Green Consumption Behavior Antecedents: Environmental Concern, Knowledge, and Beliefs. *Psychology and Marketing, 31*(5), 335–348. doi:10.1002/mar.20698

Pantelic, D., Sakal, M., & Zehetner, A. (2016). Marketing and sustainability from the perspective of future decision makers. *South African Journal of Business Management, 47*(1), 37–47.

Paul, J., Modi, A., & Patel, J. (2016). Predicting green product consumption using theory of planned behavior and reasoned action. *Journal of Retailing and Consumer Services, 29*, 123–134. doi:10.1016/j.jretconser.2015.11.006

Rashid, N. R. (2009). Awareness of eco-label in Malaysia's green marketing initiative. *International Journal of Business and Management, 4*(8), 132–150. doi:10.5539/ijbm.v4n8p132

Rex, E., & Baumann, H. (2007). Beyond ecolabels: What green marketing can learn from conventional marketing. *Journal of Cleaner Production, 15*(6), 567–576. doi:10.1016/j.jclepro.2006.05.013

Roberts, D. (2008). Thinking globally, acting locally—institutionalizing climate change at the local government level in Durban, South Africa. *Environment and Urbanization, 20*(2), 521–537. doi:10.1177/0956247808096126

Roberts, J. A. (1996). Green consumers in the 1990s: Profile and implications for advertising. *Journal of Business Research, 36*(3), 217–231. doi:10.1016/0148-2963(95)00150-6

Roe, B., Teisl, M. F., Rong, H., & Levy, A. S. (2001). Characteristic of consumer-preferred labelling policies: Experimental evidence from price and environmental disclosure for deregulated electricity services. *The Journal of Consumer Affairs, 35*(1), 1–26. doi:10.1111/j.1745-6606.2001.tb00100.x

Rousseau, G., & Venter, D. (2001). A multi-cultural investigation into consumer environmental concern. *SA Journal of Industrial Psychology, 27*(1), 1–7. doi:10.4102ajip.v27i1.768

Ryan, B. (2006). Green Consumers A Growing Market for Many Local Businesses. *UWExtension, 1*(123), 1-2.

Scott, L., & Vigar-Ellis, D. (2014). Consumer understanding, perceptions and behaviours with regard to environmentally friendly packaging in a developing nation. *International Journal of Consumer Studies*, *38*(6), 642–649. doi:10.1111/ijcs.12136

Sharma, Y. (2011). Changing consumer behaviour with respect to green marketing–a case study of consumer durables and retailing. *International Journal of Multidisciplinary Research*, *1*(4), 152–162.

Sinnappan, P., & Rahman, A. A. (2011). Antecedents of green purchasing behavior among Malaysian consumers. *International Business Management*, *5*(3), 129–139. doi:10.3923/ibm.2011.129.139

Smith, E. E., & Perks, S. (2010). A perceptual study of the impact of green practice implementation on the business functions. *Southern African Business Review*, *14*(3), 1–29.

Soyez, K. (2012). How national cultural values affect pro-environmental consumer behavior. *International Marketing Review*, *29*(6), 623–646. doi:10.1108/02651331211277973

Statistics, S. A. (2011). Msunduzi Municipanilty. *Statistics SA*. Retrieved 13 May 2017 from http://www.statssa.gov.za/?page_id=993&id=the-msunduzi-municipality

Stern, P. C. (2000). New environmental theories: Toward a coherent theory of environmentally significant behavior. *The Journal of Social Issues*, *56*(3), 407–424. doi:10.1111/0022-4537.00175

Straughan, R. D., & Roberts, J. A. (1999). Environmental segmentation alternatives: A look at green consumer behavior in the new millennium. *Journal of Consumer Marketing*, *16*(6), 558–575. doi:10.1108/07363769910297506

Strode, A., Slack, C., & Essack, Z. (2010). Child consent in South African law: Implications for researchers, service providers and policy-makers. *SAMJ: South African Medical Journal*, *100*(4), 247–249. doi:10.7196/SAMJ.3609 PMID:20459973

Synodinos, C., & Bevan-Dye, A. (2014). Determining African Generation Y Students' Likelihood of Engaging in Pro-environmental Purchasing Behaviour. *Mediterranean Journal of Social Sciences*, *5*(21), 101–110.

Takács-Sánta, A. (2007). Barriers to environmental concern. *Human Ecology Review*, *14*(1), 26.

Thøgersen, J., & Ölander, F. (2003). Spillover of environment-friendly consumer behaviour. *Journal of Environmental Psychology, 23*(3), 225–236. doi:10.1016/S0272-4944(03)00018-5

Tilikidou, I. (2007). The effects of knowledge and attitudes upon Greeks' pro-environmental purchasing behaviour. *Corporate Social Responsibility and Environmental Management, 14*(3), 121–134. doi:10.1002/csr.123

Tobler, C., Visschers, V. H., & Siegrist, M. (2012). Addressing climate change: Determinants of consumers' willingness to act and to support policy measures. *Journal of Environmental Psychology, 32*(3), 197–207. doi:10.1016/j.jenvp.2012.02.001

United Nations. (2015). *World Population Prospects: The 2015 Revision*. Retrieved 13 May 2017 from: https://populationpyramid.net/world/2020/

United Nations. (2016). *Environment Annual Report 2016*. Retrieved 13 May 2017 from: http://web.unep.org/annualreport/2016/index.php

van Riper, C. J., & Kyle, G. T. (2014). Understanding the internal processes of behavioral engagement in a national park: A latent variable path analysis of the value-belief-norm theory. *Journal of Environmental Psychology, 38*, 288–297. doi:10.1016/j.jenvp.2014.03.002

Vicente-Molina, M. A., Fernández-Sáinz, A., & Izagirre-Olaizola, J. (2013). Environmental knowledge and other variables affecting pro-environmental behaviour: Comparison of university students from emerging and advanced countries. *Journal of Cleaner Production, 61*, 130–138. doi:10.1016/j.jclepro.2013.05.015

Vikan, A., Camino, C., Biaggio, A., & Nordvik, H. (2007). Endorsement of the New Ecological Paradigm A Comparison of Two Brazilian Samples and One Norwegian Sample. *Environment and Behavior, 39*(2), 217–228. doi:10.1177/0013916506286946

Yadav, R., & Pathak, G. S. (2016). Intention to purchase organic food among young consumers: Evidences from a developing nation. *Appetite, 96*, 122–128. doi:10.1016/j.appet.2015.09.017 PMID:26386300

Young, W., Hwang, K., McDonald, S., & Oates, C. J. (2010). Sustainable consumption: Green consumer behaviour when purchasing products. *Sustainable Development, 18*(1), 20–31.

KEY TERMS AND DEFINITIONS

Eco-Labels: Symbols indicating that a product is environmentally friendly.

Environmental Concern: Consumer concern for the environment and the sustainability thereof.

Green Consumer Behavior/Ecological Conscious Consumer Behavior (ECCB)/Pro-Environmental Consumer Behavior/Environmentally Friendly Consumer Behavior: The consumer behavior of supporting green initiatives such as energy conservation as well as buying green products which do not harm the environment.

Green Consumers: People who have a particular interest in and awareness of environmental problems and make an ongoing attempt to change their behaviors to protect the environment.

Green Gap: The gap between consumers expressed environmental concern and their actual behavior.

Green Products: Environmentally friendly products that do not negatively affect the environment.

Perceived Consumer Effectiveness (PCE): The extent to which consumers believe that their individual efforts can have a positive impact on the environment.

Section 4
Sustainable Development

Chapter 6

"Airpocalypse" or Tsar Economic China:
Analysis of Unsustainable Environment and Reason Behind Increased Global Warming

Fauzia Ghani
G. C. University, Pakistan

Komal Ashraf Qureshi
Kinnaird College for Women, Pakistan

ABSTRACT

This chapter focuses on the case study of China, which is facing grave issues regarding environment and global warming. Hence, the "Airpocalypse" in China led to need and debate about the sustainability of the environment. In this chapter, an effort has been made to analyze the environmental sustainability risk which the country of China can have for the increasing rate of global warming, and how this part of region can have a transnational impact on other neighboring countries when it comes to the cause of making environment pure from pollutants, carbon dioxide, and coal emissions. The methodology of this research is qualitative, descriptive, and analytical in nature. This chapter includes the variable of environmental sustainability which is dependent on the energy consumption of industries of China involved in emission of greenhouse gases.

DOI: 10.4018/978-1-5225-3990-2.ch006

INTRODUCTION

There is a strong nexus between making money and degrading the pure environment alongside. In reality, if companies or industries would have to pay the costs they are blamed for, they would never able to get any profit from what they are making in order to attain economic benefits. This cost is nothing but a continuous damage to the environmental, which was supposed to remain pure. However, this is not the case now as the environment of the world is suffering a lot mainly because there is one to one relationship between the revenue of the corporate companies and the environmental damage. For example, if there is a drought problem in some place, it is not considered as the environmental problem only rather it is nothing but a fundamental risk to the corporations who are running their businesses because then they have to involve in the process of shifting procurement or raw materials, or their sales effort in different markets. According to the research, one of the murky reality is that the corporations are working without making any commitments for putting in their fair share to the sustainability of the environment. This fact has been observed especially in making policies to refrain from economic activities which are involved in increasing global warming rate.

This chapter has focused on the case study of China, which is facing grave issues regarding environment and global warming. The history of economic growth and boom in China although greatly increased, but has created many problems for a sustainable environment. According to the research, the devastated environmental situation in China is not the result of recent economic policies, but it is the result of the manufacturing institutions involved for over centuries. However, the country started to took some initiative when in 1972, when United Nation Organization on the Human Environmental conditions asked China to start the developmental policies by establishing the environmental institutions. There is no doubt that China owns mostly the state-owned markets and corporations but there is a need for the government to revisit the structure of the state-society relations with the bureaucratic power. China is considered to be the largest emitter of the greenhouse gases in the world according to the statistics of 2014. Moreover, the consumption of the country's energy has been ballooned immensely by late 2015. According to the Greenpeace East Asia report of December 2015, almost 80% of the 367 countries, China has failed to meet the standards for air pollution (Hatt, 2015).

As China is the world's largest producer of coal, which consists of more than half of the world's consumption. Rapid urbanization has increased the energy demands for building of new manufacturing and industrial units. According to various experts, the cities of China are facing hazardous environmental issues regarding water depletion problem and water pollution which has become the biggest environmental challenge for the country. It has also been found out that almost half of the main rivers in

China has been labelled as totally unfit for humans. The government has failed to establish the link between the consumption and the management of resources. In the past, there have been many countries in the world which have taken effective steps to control the consumption of pollutants; for example, in 1969, when the "Chuyahoga River" located in United States of America in the state of Ohio, faced the problem of thick pollutants which led to the bereaving of fish throughout, after which in the very next year, the Environmental Protection Agency was founded in America. Similarly, in 1970, the country of Japan faced the poisonous mercury spilling problem which was coming from the plastics industries, after which strict environmental laws were passed. Recently, in Beijing, the putrid thick layer of smog of January 2013 went worse than any of the previous year which led to the explosion of the public concern in China. However, no preventive immediate measures were taken by the government of China, as public themselves started to take initiatives to save their lives (Liu & Raven, 2010).

Hence, the "Airpocalypse" in China led to need and debate about the sustainability of the environment. Questions started to rise about the turning point of China because the economy of China is not only large, but resource ravening as well due to which it is overlooking the fact that there is an alarming rate of environmental issues to be focused on. The country's focus on wealth accumulation will do nothing but amplify the disastrous climatic conditions of China. This chapter has suggested the strategies from which China can decrease the rate of carbon dioxide emissions which is the main reason for magnifying global warming in the world. Moreover, it has focused on various steps which can identity the ways to cut down the level of emission of carbon dioxide at global level by highlighting its dreadful effects on the environment.

In this chapter, an effort has been made to analyze the environmental sustainability risk which the country of China can have for the increasing rate of global warming; and how this part of region can have a transnational impact on other neighboring countries when it comes to the cause of making environment pure from pollutants, carbon dioxide, and coal emissions. The various recommendations have been made to consider the possibility in which an alignment can be created between the environmental certainties and the strategies made by corporations. This chapter has focused on the possibility to implement the collective supply-chain management strategy which can be highly effective and beneficial to both the industrial corporations and environment as well e.g. this procedure involves the supply of the limited resources that should be "fit to all" whether in private enterprises or state-owned industries in order to avoid the excess consumption of burning coal or in deforestation level. The relation between the dependent variable of the environmental sustainability has been analyzed with the independent variable of pollutant emissions caused by the manufacturing industries of China. Moreover, the buffering or intervening variables has also been analyzed which are profit constraints of industries, and the lack of commitment between the corporations and environmental organizations which creates a vacuum for the industries to produce more than the given limit. Such corporations do not consider the liabilities they have on the sustainability of environment. Hence, various solutions have been proposed in this chapter after an in-depth research which includes main focus to reduce the consumption of energy levels especially for production purposes in the industries which today is the ultimate cause for increased global warming. Also, there should be widely usage of the energy saving strategies such as to use the energy-efficient resources and machineries. Moreover, to establish the renewable energy resource industries at wide level, there should be more awareness and promotion in using the technological revolutions which can help in reducing number of units in output of production process.

Hence, in order to achieve these helpful tactics and solutions, to ensure the role of participation by the public through various modes such as market based strategies which will involve direct citizens of China like information brochures, pollution discharge fee, various bans and limitations, and to bring the regime of having social gatherings in order to construct a platform for the eradication of pollutants on social based forces. People's Republic of China should regulate and monitor the energy sources and industrial places under the strict laws and regulatory system in order to attain the target of economic activities by limiting the emission of carbon dioxide and other hazardous pollutants.

Significance of the Study

There is no doubt that in the past three decades, China is the country which has accomplished significant economic development at global level. However, this rapid development has led to the rise of the constraint for increased global warming as natural environment is consistently being ignored by the country. Hence, it became necessary to analyze the factors which have become hindrance in purifying the environment specifically in China, being the world's largest emitter of pollutants mainly carbon dioxide. This chapter has examined an issue of "externalities" which played an important role in relationship with economy of China. Moreover, due to increased concern of global environmental sustainability, possible and successful solutions had been included in the study by mentioning the examples of those countries which have tried to curtail down the level of emission of carbon dioxide and their economies are also running successfully.

Nature of the Study

The nature of the study is descriptive in nature. The secondary data has been collected through the reliable sources of internet where journals and books have been considered for the study of literature review.

LITERATURE REVIEW

Mu et al. (2014) in his article professes that as compared to the other developed countries of the world, the environmental laws in the country of China were introduced late, in 1979. Nevertheless, the country has obtained recognizable achievements, but not the effective ones. These laws include the rules for preventive level of pollution, operative utilization of resources, whereas, the most important rule is the ecological protection of not only China, but its neighboring countries as well. Moreover, the globally recognized concept of sustainable environment has not been completely implemented in the country of China. There have been hindrances, gaps, and lack of coordination between the laws and obligation to follow them, irresponsibility, imperfection in the design of the whole system, moreover, being the communist state there has been an immense difficulty for the public to participate or conduct the awareness sessions, or rally to demonstrate their concern for increased level of pollutants in their country. In the future, there are prospects for the enhancement in implementing these regulations effectively, for continuous developments to mitigate and alleviate the environmental problem, with better prospects for public participation as well.

Liu and Raven (2010) state in their article, that even after the three decades of economic development, the country of China is known concurrently both global economic powerhouse, and "Airpocalypse" for the world's largest carbon dioxide emitter. As the economy of China has grown successfully providing benefits to the world, but also facing numerous challenges for the purification of environment. Moreover, the financial crisis at global level has led China to generate such infrastructure which can fulfill both the domestic demand of consumption of energy, and to be cautious for the level of emission of externalities or pollutants which is the main reason for increased global warming in the world. Even though, China is known for having the second largest total GDP (Gross Domestic Product) in the world, but it is facing trials related to environment to use energy efficient resources. Due to the massive regional size of the country, all kinds of challenges along with opportunities are having significant impact on the world.

Raven (2010) analyzes in his article the obstacles which China is facing today, given the vast size of country, being the second largest producer of coal, complexities related to public participation, and having the world's largest economy, the country has done very little in implementing policies for environmental sustainability. Moreover, this situation to prevent the level of emission will keep on getting worse, unless substantial hard work will not be considered by the government of China. These concern mainly includes to reduce the emission of greenhouse gases mainly carbon dioxide, to mitigate the local pollution created by the industrial units which is damaging the soil, water and air of the country. All these factors in turn are creating worse effect on the agricultural activities, problem of soil erosion and concern of deforestation as well to expand the infrastructure of the country.

Kapp (1997) explains in his article the issue of communism with reference to the sustainability of environment, in context with the country of China. Author explains that the degradation of the environment is having a direct relationship with the survival of humanity. However, in the country where there is limited allowance or liberty of public participation, it becomes hard for the citizens to convey their concern to the government. Moreover, this leads to the problem of sustainability of socio-economic institutions as well, where people are unable to create awareness about the hazardous steps which government takes to develop its economy but ignores the fact of survival of its own citizens. China is known for its expansion in its industrialization as compared to agricultural activities, mainly because of the concern of soil erosion, and deforestation problem. Hence, the possible solutions which can lead to purify the environment of the country at global level includes the transformation and focus on the agricultural activities as compared to industrializing the country completely, to fulfil the vacuum which has been created between the implementation of the legislative measures related to environmental sustainability and the lack of dedication by the officials, and to create public-awareness communities. It has been explored in the literature two main factors related to the environment; social and physical components which directly effects the developmental policies of the country and quality of life as well. There is a need for creating rational guiding principles for the sustainable environmental laws, along with highlighting its importance in concern to the survival of living things in this world.

Kostka (2014) explains in his article that the national leaders of China have recently showed concern to protect and purify country's environment and to change the pollution-intensive policies towards resource-efficient policies. Furthermore, this step will eventually have a significant impact on the world as well, being the world's largest economic engine. There is an immense challenge of increased level of urbanization and industrialization which are the main reason for increased level of greenhouse gas emissions in the country. Moreover, the vacuum between the central and local government is also creating a hindrance in implementing the laws related to environmental sustainability mainly because of corruption. The article explains the barriers for effective implementation of the policies and how these barriers can be removed by creating public awareness, strict regulation, by analyzing and enable reports for institutional or behavioral obstacles by the local officials, by highlighting the best possible examples of other states like those of South America, Africa, or Iceland which have become successful in reducing the level of greenhouse gas emission with tactics like introducing supply-chain management, by removing the vacuum between the governmental regulations and process of implementation. The future prospects mainly create a need for additional effective incentives and local enactment capacities for the safety of citizens.

HISTORICAL BACKGROUND

It is a widely known fact that the activities related to the production which is related to the economy of a country usually evolves the usage of energy which in turn affects the environmental conditions. The environmental affects involves the forms of pollution such as water, air, and more importantly the emission of lethal and harmful gases like carbon dioxide which is the main component for the increase in global warming issue.

Being the "factory and industrial unit of the world", China is still in the struggling phase to make the environment pollution-free. It has a large population and the resources are less to meet the needs of the citizens. Hence, that is why, it faces more pressure when it comes to meeting the economic needs of the country than any other country in the world. In 1972, the Conference of United Nations on Human Environment held in Stockholm led to the initiative in China for making the strategies related to the protection of the environment and the first Law on the Environmental Protection Trial was officially issued in the year of 1979. After that, many publications were held for the sustainable environment and since 1997, the Central Government arranges the Seminar every year, where leaders from important organizations of the country gathers to discuss the strategies in managing country's resources, environmental problems in order to propose the counter-measures for implementing the policies effectively (Matsuno, 2009).

With the passage of time, there was a constant growth in the legislative policies regarding to environmental sustainability. Pressure for implementing the strategies to protect the environment is increasing with every year in China. According to the recent report of the "Chinese Academy of Social Sciences" the city of China known as Beijing has been marked as "barely appropriate for living beings" to live here, because of polluted air, water, and traffic-congestion problem. Moreover, according to the report published by the Ministry of the Environmental Protection (MEP),

57% of the water in 198 cities of China has been considered as "hazardously bad to drink" with almost 30% of the rivers to be highly polluted. Recently in November and December 2015, Beijing suffered from the thick, brown densely smog mainly because of the air pollution which was skyrocketed to 20 times more which is 611 in Air Quality Index than the value which World Health Organization marked as safe which ranges from 0-50 in Air Quality Index. This was China's highest air pollution up till date so far in the history as mentioned in the Appendix (Figure 2) (Patricia, 2015).

There are various examples in the world laid down by the countries which have reduced their greenhouse gas emissions by creating and implementing the alternatives to use the renewable energy resources and by taking care of the natural resources in order to sustain and to live in a healthy environment. For example, in 1990, the place of Ivory Coast of Africa has reduced the greenhouse gas emission rate by 78% by working closely on the proposed program of the United Nations Environment agency. The country took some effective measures in incorporating the technological tools in order to increase the energy efficiency which did not involve the burning of component like coal. These technological tools in the industries mainly utilized the biomass fuels which is not harmful for the environment. The program also included the restriction in implementing the strategy of reforestation to reduce the level of carbon dioxide from the environment by increasing the green land areas, to increase fertility of the soil and to promote agricultural activities in the country (Tang et al., 2015).

Similarly, in Suriname which is a South American country, has also made a prominent success in creating a healthy sustainable environment in the region. Since 1990, this country reduced almost 75% of the hazardous gas emissions. Suriname is known famously for the protection of its natural resources due to which 94% of its land is forested even today which can absorb almost 340 tons of carbon dioxide gas. The country also started many projects in meeting the electrical energy requirements through hydropower stations.

Iceland also became successful in lowering the levels of gases which are responsible for increase in global warming. The nation of Iceland has become a leader in producing 100% of the electricity needs through hydropower and geothermal generators.

Moldova is another great country which has achieved the target of sustainable and pollution free environment by reducing 70% of the hazardous emission since 1990. The country has followed the project and guidelines given by the United Nations Framework Convention on the Climate Change because of which the country only contributes 0.03% of the greenhouse gases worldwide (Tang et al., 2015).

WHY THERE IS A NEED FOR ENVIRONMENTAL SUSTAINABILITY IN CHINA?

Due to the rapid increase in the urbanization and modernization, there has been continuous increase in the environmental pollutants immensely throughout the world. This has in turn led to the threat to the eruption of global warming problem which is decreasing the life expectancy rate of not only humans but also for the marine and land creatures as well. Although late 1990's and the start of the 20th century saw the important measures in the regulatory framework for the environmental sustainability management, but the monitoring and evaluation process still needs to be assessed in order to generate better results for the welfare of not only the citizens of China but also for the neighboring countries who are becoming victim of the hazardous environmental gases.

Furthermore, the agrarian culture of People's Republic of China is decades old, and the government of China is trying its best to save the legacy of the forestry lands which is helpful in fulfilling the needs of fuel, fiber, and enabling food resources. Unfortunately, because of the increase in the demand to meet the production and economic activities, there has been a drastic increase in the deforestation level. It should be noted here that the degradation of the social environment is the top-notch societal institutional, political and an ultimate economical problem in the world. China, being the largest economy generator of the world, the country possess the greater responsibility in controlling both the economic activities and environmental policies effectively in order to keep the balance in the social costs which occur during the production processes. Not only this, the colonization dependence and disintegration due to vulnerable political system made China a country with environmental dilapidation in both physical and social forms because of which in 2005, China has been remarked as the most polluted country in the world. The conflict between the environment related problem and economic developmental activities is a universal phenomenal issue worldwide and no country can escape from this (Williams, 2014).

According to the Copenhagen climate change talks which held in the year of 2009, China gained the poor repute in accelerating the hazardous greenhouses gas emitting problem in the world. Moreover, the policies which were initiated by the government to control this problem had mainly faced the dilemma of domestic suppression and businessmen considerations which would not affect their own industrial production process. As referred in the Appendix (Figure.1), it is clearly mentioned in the graph that China has been considered as the largest emitter of the carbon dioxide gas from 2000-2011 among all the other major developed countries in the world i.e. United States of America, Europe and Australia. In People's Republic of China, the theory of keeping the "Economic benefits above everything" has been the main hurdle in the implementation of the environmental sustainability goals, rather, the government of China should have focused for the "Priority on Prevention and treatment" of the problems that are polluting the environment of the country.

THREE "E'S" OF PEOPLE'S REPUBLIC OF CHINA (ECONOMY, ENERGY AND ENVIRONMENTAL CHALLENGE)

Energy

In China, coal is considered to be the main component in the process of energy consumption in almost all the industries. Its consumption is almost 60% of the total energy used in the industrial sector as calculated in 2013. China is considered to be the world's major coal manufacturer. However, due to the increase in demand in its industrial sector immensely, it has also became the largest net importer for coal as well and it ranks at the second position after the country of Australia. With the increase in the depletion of biodiverse lands due to the water pollution, there is an increase in the depletion of fossil energy (shale oil, or deep water oil) which has led to the increase in its cost as well. Hence, with the passage of time, the cost of energy resources in China is increasing due to which is importing on widely basis from the other countries (Young, 2015).

Today, capacity, consumption, cost and failures to implement the strategies to stop the depletion of natural resources in China is the major hurdle to meet the supply of energy that can be extracted from the natural resources within country, because of which the government is helpless to support the industrial sector (reason for carbon dioxide emission) in order to meet the economic demands of the country.

Economy

The economic growth of China can only be fueled if there would be abundant supply of energy resources in the country such as to meet the demand of depleted fossil fuels which are essential in extraction of oil. Rapid growth in the industrial sector of China has indeed improved the living standards of the citizens, but at the same time, it has led to the boom in their demands and concurrently polluting the environment as well due to the carbon dioxide gas emission.

Environmental Challenge

There are many challenges which the environment of China is facing nowadays; this includes the global warming due to the hazardous greenhouse gas emission worldwide in which China ranks at the top, other challenges includes the water pollution, fossil fuel depletion, noise pollution, and inability to use the renewable energy resources. Most of the air pollution in China occurs from the burning of coal in the industries which is considered to be the main cradle for 90% sulphur dioxide, 70% dust emissions, and 67% of nitro dioxide. Moreover, these hazardous gases leads to the formation of the smog as well in the air which makes difficult to breathe in the outside environment. According to the research, a country which is highly depended on the coal for its economic development is considered to be the most polluted country. Coal will still be considered to be the main component in the industrial and energy sector of China in future as well in order to maintain the status of "Tsar Economic China" in the world and to meet the high living standards of its population in today's globalized and modernized world as referred in the Appendix (Figure 10). That is why, on international platform, China still exhibits the repute of a main perpetrator when it comes to issue of environmental sustainability and to take the effective steps in decarbonize its environment. However, the world is encouraging China's policies for the construction of renewable energy industries within country, and the targets to meet the energy demands purely from these renewable energy resources by cutting down the level of using coal will be achieved in 2020 according to the 13th Five Year Plan of China (Paricia, 2015).

MAIN TYPES OF POLLUTION AND FACTORS THAT ARE IGNITING GLOBAL WARMING PROBLEM

Every year, because of the increase in pollution in the country of China, almost half a million citizens are being killed due to the hazardous environment of the country. Following are the main types of pollution which are responsible for this.

Air Pollution

The major and primary source of the air pollution in China has been the transportation vehicles which are growing at an alarming rate which is contributing in increasing the rate of Sulphur dioxide responsible for acid rain in the environment. China has declared its vehicle industries as one of the top-notch for the economic development of the country due to which the number of vehicles has also been tripled to more than 45 million in recent years. Apart from the vehicle industries, coal combustion and dust has also contributed to the air pollution of the country. These industries includes mostly the building material industries of cement and steel which are responsible for the emission of pollutant gases like nitrogen dioxide and Sulphur dioxide which accounts almost 70% of the total dust and gas emissions. The environment of China is mostly considered to be haze in majority days of the month filled with smoke and dust. The year of 2013 is considered to be the worst year when it comes to air pollution in last 50 years of China. Out of 161 cities of China, only 16 cities met the standards of the healthy air environment according to the Air Quality Index 2013 (Qin, 2015).

Water Pollution

To calculate the contamination level of water, a scale of 1-5 is used in which grade 4-5 is considered to be unfit and unhealthy to drink for humans. According to the Ministry of Land and Resources, 60% of the water in China was more than 4 grade in 2013. Not only this, 67% of the water in the China's important rivers had the contamination level range of grade 1-3. Due to this reason, more than half of the societal protests in China are related to the contaminated water and pollution (Qin, 2015).

Soil Erosion

China was known for its agrarian land decades ago, but after the National Soil Pollution Survey statistics of April 2014 under the ministries of Environmental Pollution and Land and Resources, it was found out that 16% of the soil of China has been polluted beyond the quality standardized index. According to the recent stats, the capacity of China to increase its self-sufficiency ability in production of its natural resources especially to maintain its agrarian land which was considered to be its legacy since decades is suffering immensely, due to which the country's capability to produce even grain is decreasing as well, as referred in the Appendix (Figure 9). Not only this, according to the research conducted by the Ministry of Environmental Protection from 2005-2013, 16% of the country's soil is polluted with the contaminated extracts of poisonous water and heavy-metals which is inappropriate to grow crops and it constituted of more than 65 million acres of contaminated land.

Waste From Industrial Units

With the increase in modernization and to improve the standards of living, China has excelled tremendously in the economic related activities which in turn has increased the units of industries within the country. According to the China's Statistical Yearbook on the Environment 2014, it was found out that the solid contaminated waste materials raised to 3.3 billion tons. Among these industries, steel, coal, metal, chemical and mining industries are prominent in polluting the environment which constitutes 87% of the total pollution at national level (Qin, 2015).

LAWS AND POLICIES REGARDING ENVIRONMENTAL PROTECTION

Article 26 of the constitution of China states that "the state needs to protect and take steps in protecting the environment where its people lives and this includes the ecological environment as well." At the national level, usually the strategies formulated by the State Environmental Protection Administration (SEPA formed in 1998) and the policies that are being approved by the State Council are to be implemented. The responsibilities of SEPA includes the monitoring the policies related to the environmental sustainability and to collect the relevant data in order to provide proposals at both national and international level.

The Five-Year Plans of China (First plan was constructed in 1953) which includes the country's important economic policies mainly includes the goals to decrease the energy consumption of the country relating to the economic activities. The 11th Five Year Plan (2006-2010) mainly consist of the political concern to reduce the consumption per unit of Gross Domestic Product by twenty percent (20%) and to reduce the pollutant level by ten percent (10%) as referred in the Appendix. The 11th Five Year Plan includes the restrictive marks for emission of hazardous gases like Sulphur dioxide, whereas, the 12th Five Year Plan (2011-2015) includes the other gases of carbon dioxide, nitrogen oxide, and ammonium nitrate emissions. According to the research, the monitoring and implementing equipment and criteria are still inefficient to meet the respective targets mainly because of the delays at the government end which has resulted into the weak and poorly synchronized enforcement laws in the country related to the environmental sustainability. Regardless of the delays in the monitoring of the environmental policies, China became successful in developing the largest renewable energy industrial units in the country and is considered to be the largest top country to invest into this sector especially to reduce the emission rate of hazardous gases as referred in Appendix (Figure 8). Today, it has now 20 nuclear-power reactors under operation, whereas another 28 are under the process of construction which has increased the energy capacity building of the country mostly on the renewable energy resources such as wind, solar, biomass, hydropower, or ocean (Tang et al., 2015).

The 13th Five Year Plan (2016-2020) includes the target of constructing the environmental technological industries in order to protect the ecological culture of China. Most of the policies have been taken from the "Energy development strategy action plan 2014-2020" and the "Environment protection law (2015)". The country will be focusing on commissioning 6-8 nuclear reactors every year in order to cut down its dependency on coal; this will also expand the capacity on the renewable energy resources as referred in the Appendix (Figure 10). A formal ban has been imposed on the natural deforestation due to which the amount has been reduced to 6.3%; moreover, the stress on the import of timber from Russia, Afria and Southeast Asia has also been increased in toder to fill the neccessity gaps of wood availability in the country. For water pollution, the Green tax system has been implemented in which the industrial companies are levied for paying extra taxes for polluting water, air, even on creating noise pollution, alongwith the monitoring of sewage and solid-waste inspection as well. The new Five Year Plan has also strengthened the courts and law agencies by removing much of the responsibilites from the local officials who were considered to be the main reason behind corruption with the involvement of industrial owners. Hence, the plan has also given power to the Non-governmental agencies to file the lawsuit against the polluter industries and the government has enabled more reassurance and encouragement to ensure this controlling system regarding polluting the environment (Patricia, 2015).

ANALYSIS

The variable of environmental sustainability which is dependent on the energy consumption industries of China involved in emission of greenhouse gases has been analyzed. These has been depicted in the following flowchart of the theoretical framework which shows the buffering factors working as a cushioning and catalyst forces between the independent and dependent variables. These buffering factors also increases the chances of conflict between the economic growth of the country regulated by the state-owned private enterprises and monitoring and evaluation industries responsible for the environmental sustainability of China. Hence, if the buffering or intervening factors can be mediated such as bureaucratic profits, revenues of state-owned corporations, ineffective implementation for reducing the energy demands between the dependent and independent variables, there can be effective chances for the sustainable pollutant free environment of China.

SOLUTIONS FOR CHINA TO REMOVE THE TAG OF "AIRPOCALYPSE"

After the in-depth study of the environmental related issues of China, an effort has been made in explaining that why China is forced to allocate insufficient resources and strict policies to control the coal emission rate in manufacturing industries; mainly because the country is being pressurized to meet both the needs of its citizens and to generate abundant economic activities in order to consider as the world's largest economic hub. However, this can impact the economic activities of the country, as it has to control the energy demands as well, and China has taken sufficient steps and made many legislative policies to answer the International platform regarding its control on the greenhouse gas emissions. In this chapter, this very critical position of China has been described for having the largest economic hub in the world, and becoming a threat to the global environmental sustainability alongside due to which many countries especially western countries are criticizing China for not taking eminent steps to control the pollutant situation of the country. Following possible initiatives can be taken by China for making the better environment:

Green-Economy Production

Although China has taken many steps towards the green-economy (an economy which focuses on the well-being and better standards of human lives, reducing inequalities, and where there is an effort made to remove the ecological problems), but the implication procedure still suffers from numerous challenges especially when it comes to the investment process. There is a consistent lack in the implementation procedures of the legislative policies such as poor evaluation system, there is a pressure on the local officials to impose high tax rates or to charge the industrial owners for polluting the environment. In this very procedure, these officials are often being bribed by these owners, which creates the loose-end into the implementation of the policies. Central government still plays a vital role which should make sure of the transparency at all the hierarchical levels in the system. Businesses are consider to be the key component in the green-economy, for example, in renewable energy resources, in the construction of the eco-friendly environment, in promoting the concept of green-transportation industries. There is a dire need to distinguish between the state-owned corporations and the small-scale companies that can create great impact in the sustainable environment of China. There is a need for a cooperation and friendly synchronization between these two types of enterprises in order to implement effective strategies in the country.

Public Participation in Environmental Problems of China

There is a need for the government of China to initiate the social gatherings in which various Non-governmental organizations should also participate to promote the awareness regarding the environmental sustainability problems in the country. In the year of 2006, the first measure regarding the public participation in Environmental Impact Assessment was formulated in which it was made essential to involve social groups of people and related enterprises in the expansion of environmental related information which can help the citizens in enabling them guidance and support to make the environment of the country pollution-free. By creating the communities which can help in making the environment healthy and which can create awareness is one of the effective and innovative way to achieve the respective targets in reducing the pollutants.

Collective Supply-Chain Management Strategy "Fit to All"

This process can be very effective in controlling, monitoring, and limiting the usage of resources in both the private and state-owned corporations. For example, recently, according to research, almost 80% of the world's industries of palm oil have made the commitment to follow the sustainable practices in which the industries fulfill specific demands of the commodity producers and do not go for excess production; due to which a significant amount of palm oil manufacturers have contributed in decreasing of the deforestation rate. This will decrease the burning rate of the coal in the industries, and will meet the economic and public demands of the country concurrently as well according to the planned target. This can only be initiated under the strict evaluation systems of central government, in order to control the resources of the country and to allocate the revenues generated from the economic activities appropriately. There is a need for the government and businessmen to realize some costs by keeping their self-interest aside in order to implement the strategies for cleaning of environment by staking some risk in country's economic growth to control the emission level of hazardous greenhouse gases.

FUTURE RESEARCH DIRECTIONS

There is a need for analyzing the factors which are becoming hindrance in implementing the environmental sustainability laws and regulations in China. Various research questions need to be further study in-depth such as: What are the guiding principles and sustainability goals of China to provide a rationale for improving its environmental conditions? How such priorities can be established and evaluated effectively keeping in context the escapes created by the local officials for ignoring the laws? What kind of role public participation can play for ensuring a sustainable environment of their country? What kind of relevance is there between the traditional Chinese attitudes towards its nature and contemporary need of purifying the environment? For how long the present approach of China for creating a sustainable environment can last and what is its possible validity which can implemented on other developing or developed countries?

CONCLUSION

Today, the central government of China realizes the importance of clean and healthy environment because of which it has initiated the strategies of heavy penalties, taxes or subsidies to those industrial units or enterprises which pollutes the environment, but such strategies can also lower the interest of the workers and producers as they will not be able to meet their required targets and they can face loss as well. Hence, a set of more comprehensive rules are still needs to be established when it comes to the imposition of bans or high-rate taxes. There is a need for the central government to cooperate with the local officials in order to remove the hierarchical barriers for proper implementation of the environmental policies. There is a need for the promotion of problems which are related to the environment in the form of societal groups, in the educational institutions and with the help of NGO's as well. If the government of China try to shift the structural hierarchy towards decentralization, there can be a prosperous rate in reducing the pollutant level in order to effectively implement environmental sustainability policies in the country. The tag of China to be known now as the "Airpocalypse" rather to be called as "Tsar of economic activities" is basically a sign to have a radical and prominent change in the methods of economic controlling activities; also in the structure of rigid governmental hierarchy. Hence, in order to remove this tag, it is essential to make its environment "pollutant free". There is a dire need for the country to realize the modification challenges in creating a balance between the ecological environment and economic development of the country in which the main stake is nothing but the lives of country's citizens.

REFERENCES

Adams, P. (2015). *The truth about China; Why Beijing will resist demands for abatement.* The Global Warming Policy Foundation.

Adams, P. (2015). The truth about China; Why Beijing will resist demands for abatement. *The Global Warming Policy Foundation GWPF Report 19*.

Anders, E. (2008). *China environment and climate change policy brief.* University of Gothenburg.

Andersson, L. (2006). *China's environmental policy process: A case study of the challenges of creating and implementing air pollution policies in China.* University of Lund, Department of Political Science.

Baeumler, A., Ijjasz-Vasquez, E., & Mehndiratta, S. (2012). *Sustainable low-carbon city development in China.* The World Bank.

Canfa, W. (2007). Chinese environmental law enforcement: Current deficiencies and suggested reforms. *Vermont. Journal of Environmental Law, 8*, 159–190.

Chow, G. C. (2007). China's energy and environmental problems and policies. *CEPS Working Paper, 152*, 1-18.

Freeman, C. W., & Lu, X. (2008). *Assessing Chinese government response to the challenge of environment and health.* Center for Strategic and International Studies.

Hatt, V. (2015, January 13). *Readers Supported News.* Retrieved from http://readersupportednews.org/opinion2/271-38/28024-global-warming-is-a-national-security-threat

Kapp, W. (1997). Environmental policies and development planning in contemporary China. The Hague, The Netherlands: Academic Press.

Kostka, G. (2014). Barriers to the implementation of environmental policies at the local level in China. *Policy Research Working Paper, 7016*, 1-59.

Kunmin, Z., Zongguo, W., & Liying, P. (2007). Environmental policies in China; Evolvement, features, and evaluation. China Population, Resources and Environment, 17(2), 1-7.

Liu, J., & Raven, P. H. (2010). China's environmental challenges and implications for the world. *Critical Reviews in Environmental Science and Technology*, 1–29.

Man, J. Y. (2013). *China's environmental policy and urban development.* Lincoln Institute of Land Policy.

Marks, P. (2005, June 8). *News Scientist.* Retrieved from www.newscientist.com: https://www.newscientist.com/article/dn7488-global-warming-is-a-clear-and-increasing-threat/

Matsuno, H. (2009). *China's environmental policy: Its effectiveness and suggested approaches for Japanese companies.* Nomura Research Institute.

Mu, Z., Bu, S., & Xue, B. (2014). Environmental legislation in China; Achievements, challenges and trends. *Sustainability*, 1–13.

OECD. (2006). *Environmental performance review of China.* OECD.

Qi, Y. (2007). *China's Challenges in environmental regulation.* School of Public Policy and Management: Tsinghua University.

Qin, X. W. (2015). *China's path to a green economy: Decoding China's green economy concepts and policies.* International Institute for Environment and Development.

Rozelle, S., Huang, J., & Zhang, L. (1997). Poverty, population and environmental degradation in China. *Food Policy*, 22(3), 229–251.

Kin, S. K., Hao, R., & Champeau, A. (2012). *Growth, Environment, and Politics; The case study of China. Conference on China*, Seattle, WA.

Tang, X., McLellan, B. C., Snowden, S., & Hook, M. (2015). Dilemmas for China; Energy, Economy and Environment. *Sustainability*, 1-13.

WHO. (2001). *Environment and People's health in China.* United Nations Development Programme.

Williams, L. (2014). *China's climate change policies.* Lowy Institute.

Wu, J., Deng, Y., Huang, J., Morck, R., & Yeung, B. (2013). *Incentives and outcomes; China's environmental policy.* National University of Singapore.

Wu, T. (2011, November 7). *The Diplomat.* Retrieved from www.thediplomat.com: http://thediplomat.com/2011/11/china-brics-and-the-environment/

Xu, H. (1998). Environmental policy and rural industrial development in China. *Research in Human Ecology*, 6(2), 72–80.

Young, A. (2015, March 22). *IB TImes.* Retrieved from www.ibtimes.com: http://www.ibtimes.com/global-warming-china-wealth-accumulation-will-amplify-climate-disasters-china-says-1855000

Zhang, Q., Crooks, R., & Jiang, Y. (2011). *Environmentally sustainable development in the People's Republic of China; Vision for the future and the rold of the Asian Development Bank.* Asian Development Bank.

Zhu, J., Yan, Y., & He, C. (2015). *China's environment; Big issues, accelerating effort, ample opportunities.* Equity Research.

Zhu, J., Yan, Y., He, C., & Wang, C. (2015). *China's environment; Big issues, accelerating effort, ample opportunities.* Goldman Sachs: Equity Research.

US Energy Information Administration. (2013). Total Carbon Dioxide Emissions from the Consumption of Energy (Million Metric Tons). *International Energy Outlook 2013.*

AirNow. (n.d.). *Current AQUI.* Retrieved from https://www.airnow.gov

Frankfurt School-UNEP Centre/BNEF. (2014). *Global Trends in Renewable Energy Investment 2014.* Retrieved from http://fs-unep-centre.org/publications/gtr-2014

REN21. (2014). *Renewables 2014 Global Status Report.* Retrieved from http://www.ren21.net/Portals/0/documents/Resources/GSR/2014/GSR2014_full%20report_low%20res.pdf

Xinhua News Agency. (n.d.). *Xinhua Net.* Retrieved from http://www.xinhuanet.com/english/

OECD. (n.d.). *Country Statistical Profile: China.* Retrieved from http://www.oecd-ilibrary.org/economics/country-statistical-profile-china_csp-chn-table-en

Lewis, J. (2011). *Energy and Climate Goals of China's 12th Five-Year Plan.* Retrieved from http://www.c2es.org/international/key-country-policies/china/energy-climate-goals-twelfthfive-Year-plan

KEY TERMS AND DEFINITIONS

Environmental Sustainability: To control the renewable resources, pollution, to control the depletion of non-renewable resources for an indefinite period in order to create a healthy environment for people in the world.

Greenhouse Gases: Gases that trap the heat in the atmosphere are known as greenhouse gases. These gases ultimately make the Earth a warmer place to live and lead to global warming, depletion of resources, and reduce the survival rate of living things. These gases are capable of absorbing the heat and infrared radiation in the atmosphere, which are harmful to living things on Earth, for example, carbon dioxide, methane, and ozone. Today, the increased level of industrialization, which consumes large amounts of coal, leads to greater emission of carbon dioxide, which leads to an unsustainable environment and makes it harder to live.

Non-Renewable Resources: Includes resources that can be found inside the earth, for example, natural gas, fossil fuels, and coal.

Renewable Resources: Resources that can be used repeatedly and can be replaced naturally as well. For example, oxygen, solar energy, or biomass. Such resources also include the commodities like wood, leather, and paper.

Sustainability: Considered to be the ability to pursue a defined behavior, policies for an indefinite period.

APPENDIX

For a more accurate representation of these figures see the electronic version.

Figure 1.
Source: US Energy Information Administration (2013)

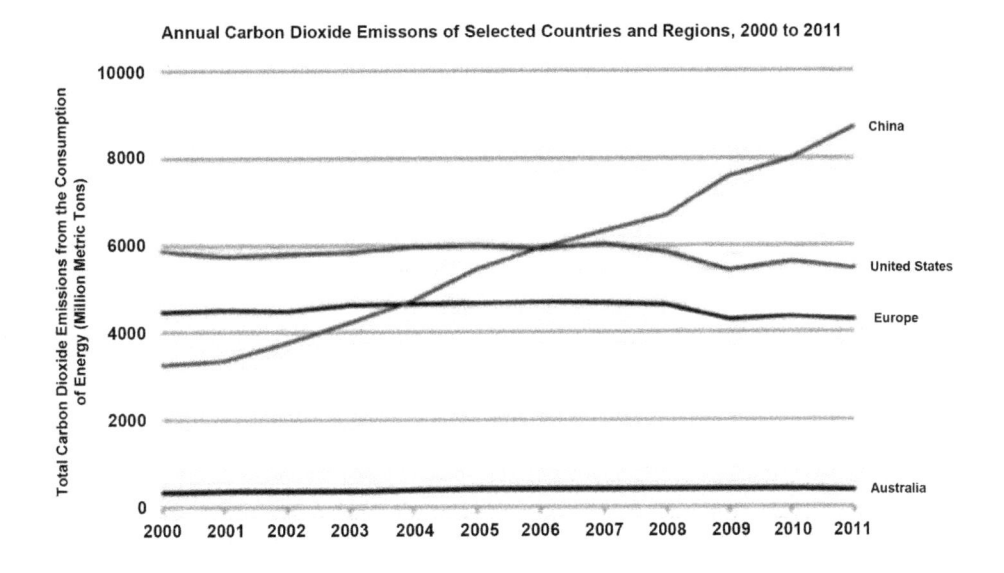

Annual Carbon Dioxide Emissons of Selected Countries and Regions, 2000 to 2011

Figure 2.
Source: Beijing's air quality index (AQI) was 611 on Monday (AirNow, n.d.)

When the AQI is in this range:	..air quality conditions are:	...as symbolized by this color:
0-50	Good	Green
51-100	Moderate	Yellow
101-150	Unhealthy for Sensitive Groups	Orange
151 to 200	Unhealthy	Red
201 to 300	Very Unhealthy	Purple
301 to 500	Hazardous	Maroon

Figure 3.
Source: Frankfurt School-UNEP Centre/BNEF (2014).

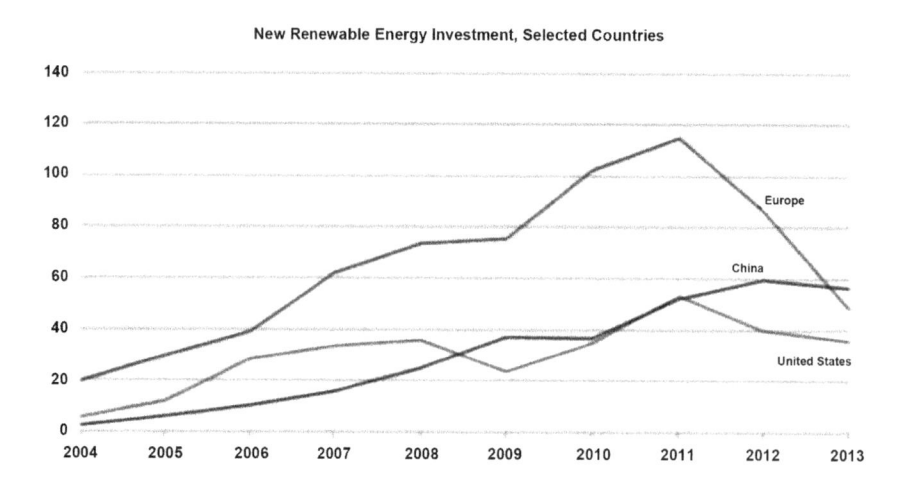

New Renewable Energy Investment, Selected Countries

Figure 4.
Source: REN21 (2014).

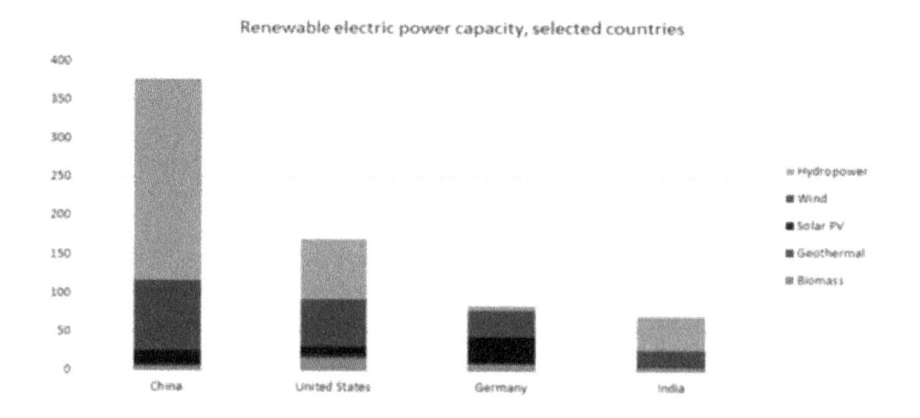

Renewable electric power capacity, selected countries

Figure 5.
Source: Xinhua News Agency (n.d.)

	12th Five-Year Plan's Targets (Compared to 2010)	12th Five-Year Plan's Achievements (Compared to 2010)	13th Five-Year Plan's Targets (Compared to 2015)
Energy Intensity (Energy Consumption per Unit of GDP)	-16%	-18.2%	-15%
Carbon Intensity (Carbon Emissions per Unit of GDP)	-17%	-20%	-18%
Non-Fossil Fuel Percentage	11.4%	12%	15%
Sulfur Dioxide (SO$_2$)	-8%	-18%	-15%
Nitrogen Oxides (NO$_x$)	-8%	-18.6%	-15%
Ammonia Nitrogen	-10%	-13%	-10%
Chemical Oxygen Demand	-10%	-12.9%	-10%

Figure 6.

Five-year plan select economic development targets	12th five-year plan target (2011-15)	2011-15 (actual)	13th five-year plan target (2016-20)
Nominal GDP	Rmb55.8trn	Rmb67.7trn	>Rmb92.7trn
Annual real GDP growth	7%	7.8%	>6.5%
Tertiary sector as a proportion of GDP	47%	50.5%	56%
Urbanisation rate	51.5%	56.1%	>60%
Urban household registration (*hukou*) rate	n/a	39.9%	45%
Urban job creation	45m	64m	50m
Construction of affordable housing units	36m	27m	n/a
R&D spending as a proportion of GDP	2.2%	2.1%	2.5%
Reduction in carbon emissions per unit of GDP	17%	20%	18%
Reduction in energy consumption per unit of GDP	16%	18.2%	15%
Farmland reserves (mu)	1.82bn	1.86bn	1.86bn
Forest coverage rate	21.66%	21.66%	23.04%

Note. 1 mu is equivalent to 666.7 sq metres
Source: Xinhua News Agency.

Figure 7.
Source: OECD (2014)

Chinese GDP Per Capita and Energy Intensity, 2006-10

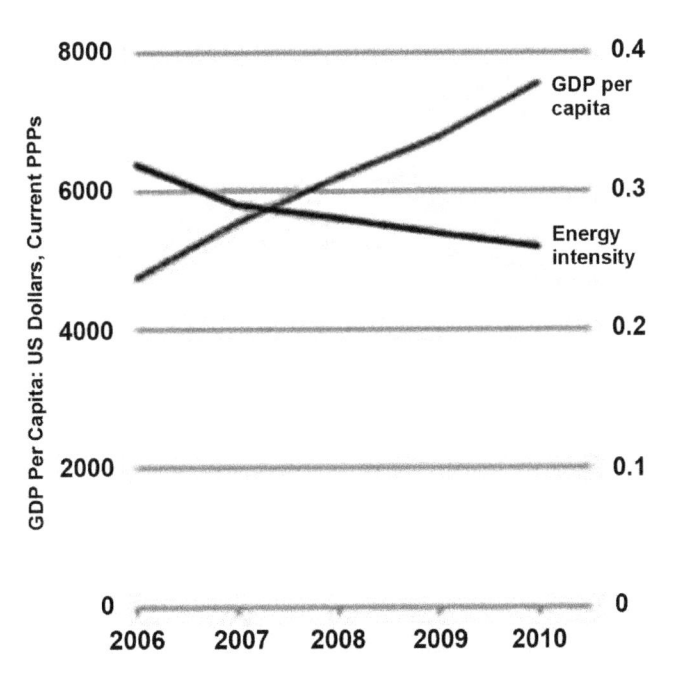

Figure 8.
Source: Lewis (2011)

12th Five Year Plan: main climate change-relevant targets

Carbon intensity (tons carbon dioxide/10,000 RMB)	Reduction of 17 per cent from 2010 levels by 2015
Energy intensity (tons standard coal equivalent/10,000 RMB)	Reduction of 16 per cent from 2010 levels by 2015
'Non-fossil' (renewable and nuclear) energy (proportion of total primary energy)	Increase from 8.6 per cent in 2010 to 11.4 per cent by 2015

Figure 9.

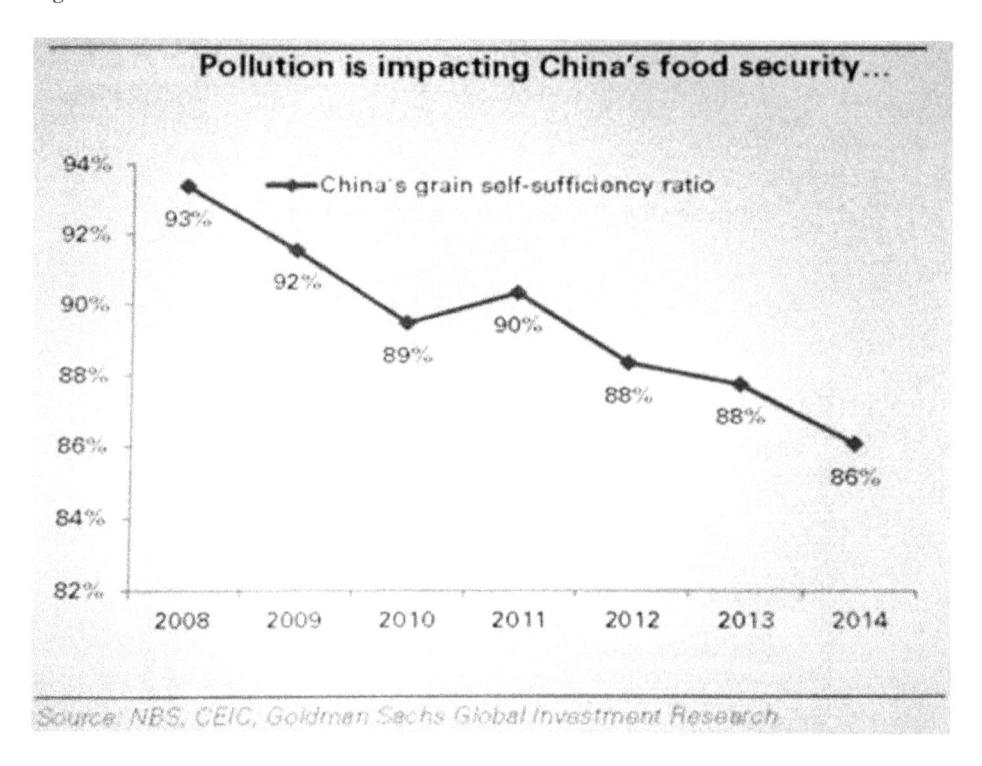

Pollution is impacting China's food security...

China's grain self-sufficiency ratio:
- 2008: 93%
- 2009: 92%
- 2010: 89%
- 2011: 90%
- 2012: 88%
- 2013: 88%
- 2014: 86%

Source: NBS, CEIC, Goldman Sachs Global Investment Research.

Figure 10.

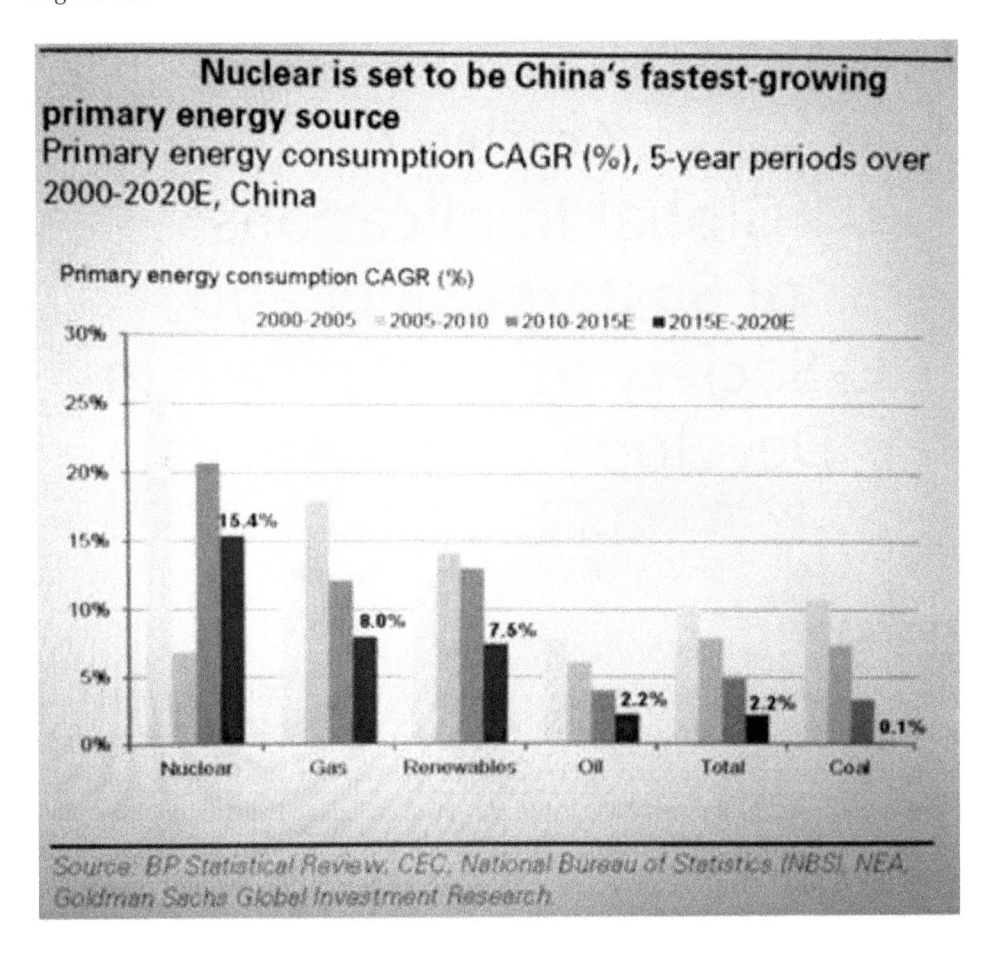

Chapter 7
Global Implications of Sustainability and E-Society Infrastructure in Developing Economies

Birău Ramona
Constantin Brâncusi University, Romania

ABSTRACT

The main purpose of this chapter is to investigate the global implications of sustainability and e-society infrastructure in developing economies. A generally accepted definition of e-society is very difficult to compress into words considering the complexity of the phenomenon itself. However, an exhaustive approach includes a great variety of original views and individual multifaceted opinions which converge to obtain a solid theoretical structure. Globalization is the modern term used to describe changes in the structure of societies and the world economy, but having a major impact in the context of an accelerated informatization. The process of globalization is not a new and innovative process, but it is the result of changes in the world economy that have increased in recent years, considering the fact that it brings a number of advantages. Moreover, globalization means labor mobility without constraint of geographical boundaries. Generally, the progress of communications is another consequence of globalization and the impact of change is even stronger in developing economies.

INTRODUCTION

Globalization is not a reversible process, but it can be perceived as a favorable phenomenon with profound social and economic influences. The economic interests generate at the same time serious social contrasts due to open borders

DOI: 10.4018/978-1-5225-3990-2.ch007

and liberalization of markets. Current trends of the economy include issues such as financial integration and liberalization but on a global scale. Unrestricted and unlimited access to broadband internet networks and multimedia communications systems raise a fundamental dimension of the digital age. Effectively, sustainability and E-Society infrastructure in developing economies cover a large field of academic and practical interest. A complex theoretical framework is required in order to define sustainability and its multiple socio-economic implications. The *concept of sustainable development* implies a very wide applicability. The major challenges of sustainable development is to achieve an implementation across government authorities, a wide range of civil society and other internal and external stakeholders. The conceptual approach of sustainable development is used in most areas of activity in order to achieve an optimal level of accountability on essential issues related to quality of life. Sustainability is intended to provide an optimal solution to negative phenomena such as massive exploitation of natural resources and the continuous degradation of natural environment. Sustainable development in the case of developing economies involves a balance between the needs of social systems and natural systems, as well as between the needs of present and future generations, according to the environmental preservation as it is undeniable globally unequal distribution of resources. On the other hand, in terms of private capital, large companies are investing heavily in other countries, including developing economies, thus becoming multinational corporations. *Developing economy* describes the economy of an emerging country and it is characterized by a lower standard of living, a lower Human Development Index (HDI) and a lower quality of life. Moreover, developing or *emerging countries* are characterized by an increased level of poverty, increased financial volatility, significant social inequity, a low level of industrialization, a fragile infrastructure, modest architecture of the health system, high fragmentation, a higher birth rate, inadequate education systems, a higher infant mortality rate.

In the context of globalization, multinational corporations arise because facilitate the transfer of technology and digitization of economies. However, it is important to reveal the fact that capital is significantly more mobile than labor. The real world and the virtual world are two realities separated by increasingly thinner line of, by reference to a digitized society. While digital boundaries become increasingly thinner, the social adaptation must fill the need of interaction between members of an E-Society. In this regard, sustainability highlights the to maintain a state of progress in the context of developing economies. Moreover, sustainability and E-Society infrastructure represent the fundamental elements in a very volatile framework given the very heterogeneous influencial factors. However, this phenomenon leads to a polarization of societies in rich societies and poor societies.

The *concept of E-Society infrastructure* highlights the very competitive and technological nature of modern day perspective. The E-Society infrastructure is based on a number of communication and technical devices in order to develop a sustainable environment. Practically, the E-Society infrastructure is based on mobile phone devices such as iPads, iPods, handheld tablet, notebook, mobile phones, mobile PC, smartphone and other portable accessories. Moreover, essential components of E-Society infrastructure include: wireless technology, social media technology (networking), satellite digital services, computer systems technology, multimedia technologies, information technology and others. An overview of E-Society infrastructure highlights the major role played by virtual reality (VR) technologies. In other words, the infrastructure provides essential support in order to make it possible to experience anything and anywhere. Nevertheless, the particularities of developing economies lead to a more realistic approach on long-term.

This chapter intended to provide a broad overview on the concept of sustainable development and its implication. Extrapolating the basic concept, we can estimate that sustainable development in emerging countries provides a balance between the needs of social systems and natural systems, between the needs of present and future generations, according to the environmental preservation considering it is a globally undeniable unequal distribution of resources. Moreover, the systemic risk caused by the increasing scarcity of natural resources must be also considered in the context of long-time horizons due to the negative impact caused by the depletion of natural resources. In the recent past, the relationship between natural resources and economic development has become extremely strong with significant implications for present and future generations.

In order to reduce the gaps between theory and practice with regards sustainable development is necessary to explore several issues, including population growth, assessing damage and benefits to the environment, poverty, social inequity and natural resource scarcity. Obviously, the heterogeneity of influencial factors is undisputable especially in terms of globalization. Moreover, it is essential to consider a long-term approach focused on the effects on present and future generations. Sustainable development concepts are increasingly used in most areas of expertise.

THEORETICAL FRAMEWORK

The main objective of this theoretical framework is to provide a clear understanding of sustainability and E-Society infrastructure in developing economies. An insight into the depth of these relatively recent and high impact concepts, suggests the need for complex and interdisciplinary approaches. A theoretical conceptualization of sustainability and E-Society infrastructure can lead to efficiency in these areas in the context of in developing economies. Over the years, it was highlighted the need for sustainable growth to rely on a strong equilibrium between economic growth and preservation of natural resources.

According to Magoulas, Lepouras and Vassilakis (2007), the concept of E-Society includes a very extensive range "of applications from e-government, e-democracy, and e-business to e-learning and e-health." Raiyn (2011) investigated the idea of developing E-Society cognitive platform and suggested that "E-Society is a general term that includes different technology resources as e-learning, e-commerce, e-medicine, and e-government that are essential for human." Schmid, Stanoevska and Tschammer (2013) investigated innovative trends in communications systems, information processing, security and trust in electronic commerce, business, and government in order to provide a more comprehensive framework on E-Society. On the other hand, Magoulas, Lepouras and Vassilakis (2007) investigated the issue of virtual reality in the E-Society and suggested that "it appears that E-Society can significantly benefit from the application of virtual reality technologies".

Cheever and Dernbach (2015) consider that sustainable development represents "a decision-making framework for maintaining and achieving human well-being, both in the present and into the future which requires both consideration and achievement of environmental protection, social justice and economic development." Rosa (2009) investigated fundamental issues on sustainability and infrastructure resource allocation and also stated that "sustainable resource use implies a concern for intergenerational equity in the long-term decision making of society".

Tan (2014) investigated challenges and issues of sustainable development policies and stated that: "green growth strategy ensures that natural resources realize its full economic potential and is maintained by regular investments." Wehrli (2014) revealed that mountains are extremely important for sustainable development because they provide goods and services but on the other hand "they are among the most disadvantaged regions in the world, with the highest poverty rates and some of the greatest vulnerability to global climatic, environmental, and socioeconomic change and related risks". Castro (2004) suggested that "like democracy and globalization, the concept of sustainable development has become one of the most ubiquitous, contested, and indispensable concepts of our time." Dale and Newman (2005) investigated a wide range of issues regarding sustainable development education representing a counterweight to traditional environmental education.

Ogbodo (2010) explores the concept of sustainable development in light of two antagonistic schools of thought highlighting that "…while the Positivists / Optimists insist on promoting development for the mutual benefit of both the present and future generations; the Negativists / Pessimists insist on restricting development in the same name of sustainable development." According to Leuenberger and Wakin (2007) the concept of sustainability represents "a philosophy associated with the long-run maintenance or improvement of human welfare and the preservation of natural capital and environmental integrity". Bina (2013) analyzed the concept of sustainable development in relation to green economy without ignoring the global implications of present reality, aspect that emerges from next statement: "reality is still dominated by concepts of development and growth, largely free of adjectives such as 'sustainable', 'inclusive', or 'green'".

Sustainable development policy aims to achieve a continuous improvement in citizens' quality of life and wellbeing, which involves the pursuit of economic progress while safeguarding the natural environment and promoting social justice (European Union – EUROSTAT, 2015).

Howard and Wheeler (2015) investigated the manner in which community development and citizen participation should contribute to the new global framework for sustainable development in order to "transform the current aid architecture and promote environmental, economic and social well-being on a global scale". George (2007) provided an interesting perspective on sustainable development and global governance, suggesting that: "the difficulties of achieving sustainable development reflect several internal tensions in the three-pillar approach…" and "…no distinction is drawn between the development of developing countries and the development of developed ones".

Hopwood, Mellor and O'Brien (2005) provided an interesting perspective about sustainable development considering that "it needs more clarity of meaning, concentrating on sustainable livelihoods and well-being rather than well-having, and long term environmental sustainability, which requires a strong basis in principles that link the social and environmental to human equity". Birău (2015) provided a statistical analysis on poverty in European Union based on a case study for Romania in order to achieve "greater convergence, sustainable economic growth and social cohesion". Awasthi (2011) investigated complex issues regarding socio-economic challenges and sustainable development in developing countries and mentioned his opinion, ie "…the developing countries have been condemned for their incapability to diminish poverty related scarcity and contribution to sustainable agricultural development."

El-Diraby and Wang (2005) discussed the idea of an E-Society portal regarding integrating urban highway construction projects into the knowledge city concidering the major importance of community involvement. Longley and Singleton (2009) provided a nationwide geodemographic classification of how people engage with new information and communication technologies (ICTs) based on public views of the geography of the E-Society. Zheng and Walsham (2008) analyzed issues of significant importance on social exclusion in the E-Society as capability deprivation considering socio-political, cultural and institutional factors as well as the effective use of information communication technologies. Raiyn (2011) suggested that "E-Society is based on the agent technologies while the social agents offer impressive, meaningful and several features as autonomy, manage negotiation and make decision".

Newman (2016) highlighted the concept of sustainable urbanization based on four stages of infrastructure planning and progress, ie the Modernist from 1940's to 1980's (phase 1), the Post-Modernist stage from 1980's to 2000's (phase 2), the emerging sustainability from 2000's to today (phase 3) and the disruptive innovation - present day to immediate future (phase 4). As a counterweight, Ugwu (2013) provided an interesting approach on an underpinning technology such as nanotechnology for sustainable infrastructure considering that "in the civil engineering and construction sector, the Brundtland report heralded a paradigm shift from traditional infrastructure procurement and delivery with the focus now on sustainable infrastructure systems". Birău (2017) investigated theoretical issues on sustainable development in public administration in Romania and argued that is mainly focused on "the optimization of the social system as a whole by focusing on public interest and ensure quality public services for citizens".

Swilling (2006) discussed relevant aspects regarding the the importance of considering ecological sustainability issues in any city's infrastructure plans and investments based on a Cape Town (South Africa) case study. Dasgupta and Tam (2005) revealed that "physically implementing sustainable infrastructure involves three life stages for any project: preproject planning, project implementation and ongoing operations", but a rigorous evaluation of sustainability in in each of these this stages is essential. Otegbulu and Adewunmi (2009) conducted a series of issues on evaluating the sustainability of urban housing development in Nigeria through innovative infrastructure management namely considering the idea of sustainable housing development. Newman (2015) investigated various aspects of great impact on transport infrastructure and sustainability based on their importance regarding economic development, planning and assessment process in order to achieve goals.

Bocchini, Frangopol, Ummenhofer and Zinke (2014) exhaustively investigated issues on resilience and sustainability of civil infrastructure, but considering that these are complementary and it is important to be used in an integrated manner mainly due to the fact that "resilience is usually connected to the occurrence of extreme events during the life cycle of structures and infrastructures". Shen, Wu and Zhang (2011) revealed significant aspects on key assessment indicators (KAIs) for the sustainability performance of infrastructure projects and predicting that "infrastructure projects will continue to be developed in the coming years, particularly in developing countries such as China and India…". Pour-Ghaz (2013) provided a very documented insight on sustainable infrastructure materials as to highlight the importance of "developing new low carbon footprint construction materials to lower the environmental emissions and implications of infrastructure". Additionally, considering the interdisciplinary aspect of sustainability, Kaminsky and Javernick –

Will (2014) analyzed significant implications regarding internal social sustainability of sanitation infrastructure.

On the other hand, Wessels (2014) believes that the inclusion of "sustainability objectives to infrastructure projects is still seen by many as an increased risk, rather than a value adding enterprise". Siew, Balatbat and Carmichael (2016) investigated a number of issues on assessing the sustainability of infrastructure and reveal that "the development of sustainability assessment and sustainability reporting tools (SRTs) for infrastructure is pivotal because they serve to inform on progress towards achieving sustainability goals".

ANALYZING THE CONCEPT OF DEVELOPING ECONOMIES

Emerging countries are the countries whose economies are in a fast increase process, respective in transition phase to a market economy (Simon, 1997). In other words, emerging countries have a higher capacity than the developed countries to provide investors opportunities in order to obtain higher profits. In terms of strategic financial management, international investors are informed of the fact that they cannot earn high profits as investing in mature markets compared to those earned by investing in emerging markets of developing countries. Moreover, in order to obtain higher returns, international investors are assuming additional risks as volatility risk, liquidity risk or currency risk, which are significantly higher than in developed markets. The ability of international investors to improve risk-adjusted returns, however, is based on the fact that benefits only target the systematic risk. Emerging markets are characterized by a high volatility of the exchange rate which obviously lead to a greater risk in financial trading.

Tehnically, emerging countries are characterized by certain stylized facts such as attracting new technologies, growth opportunities and foreign investment. Momaya (2011) investigated cooperation for competitiveness of emerging countries in order to explore a cooperation - driven stage in the traditional framework of country competitiveness development considering strategies to leverage cooperation on relevant dimensions. Mody (2004) considered that the combination of high volatility and the transitional features of emerging economies generate a real challenge in policymaking which is the appropriate balance between commitment and flexibility, or between rules and discretion. However, it is very important to understand the very nature of emerging countries in order to manage its high prospects. The transition from an emerging economy to a developed economy is a difficult and complicated process but the potential of achieving the objective is highly beneficial.

Emerging markets are much more exposed to volatile currency swings, high volatility, domestic policy instability, leveraged investments, rapid economic growth,

Global Implications of Sustainability and E-Society Infrastructure

accelerated industrialization, disaster vulnerability, higher than average financial returns, lack of effective disclosure and transparency, fragile macroeconomic stability, fiscal deficits, inadequate legal framework, social and environmental vulnerability. Across emerging markets the basic idea of sustainable development gains different valences based on the influence of transitional economy. In other words, there are some extremely varied features which distinguishes emerging markets from developed markets.

For developing economies, digitization is a long-term perspective. According to FTSE Annual Country Classification Review released in September 2015, there are the following four categories of countries, ie developed, advanced emerging, secondary emerging and frontier. Developed countries include the following countries mentioned in alphabetical order: Australia, Austria, Belgium/Luxembourg, Canada, Denmark, Finland, France, Germany, Greece, Hong Kong, Ireland, Israel, Italy, Japan, Netherlands, New Zealand, Norway, Portugal, Singapore, South Korea, Spain, Sweden, Switzerland, UK and USA. Advanced emerging countries include: Brazil, Czech Republic, Hungary, Malaysia, Mexico, Poland, South Africa, Taiwan, Thailand and Turkey. Secondary emerging countries include: Chile, China, Colombia, Egypt, India, Indonesia, Pakistan, Peru, Philippines, Russia and UAE. Frontier countries include: Bahrain, Bangladesh, Botswana, Bulgaria, Côte d'Ivoire, Croatia, Cyprus, Estonia, Ghana, Jordan, Kenya, Lithuania, Macedonia, Malta, Mauritius, Morocco, Nigeria, Oman, Qatar, Romania, Serbia, Slovakia, Slovenia, Sri Lanka, Tunisia and Vietnam.

According to the official website of the World Bank, for the current 2016 fiscal year, low-income economies are defined as those with a GNI per capita, calculated using the World Bank Atlas method, of $1,045 or less; middle-income economies are those with a GNI per capita of more than $1,045 but less than $12,736; high-income economies are those with a GNI per capita of $12,736 or more. Lower-middle-income and upper-middle-income economies are separated at a GNI per capita of $4,125. Moreover, the World Bank suggests that term country, used interchangeably with economy, does not imply political independence but refers to any territory for which authorities report separate social or economic statistics.

The category of *low-income economies* ($1,045 or less) include the following countries: Afghanistan, Gambia, Niger, Benin, Guinea, Rwanda, Burkina Faso, Guinea-Bisau, Sierra Leone, Burundi, Haiti, Somalia, Cambodia, Dem Rep. Korea, South Sudan, Central African Republic, Liberia, Tanzania, Chad, Madagascar, Togo, Comoros, Malawi, Uganda, Congo, Dem. Rep Mali, Zimbabwe, Eritrea, Mozambique, Ethiopia, Nepal. The category of *lower-middle-income economies* ($1,046 to $4,125) include the following countries: Armenia, Bangladesh, Bhutan, Bolivia, Cabo Verde, Cameroon, Rep. Congo, Côte d'Ivoire, Djibouti, Arab Rep. of Egypt, El Salvador, Georgia, Ghana, Guatemala, Guyana, Honduras, India, Indonesia, Kenya, Kiribati,

Kosovo, Kyrgyz Republic, Lao PDR, Lesotho, Mauritania, Micronesia, Moldova, Morocco, Myanmar, Nicaragua, Nigeria, Pakistan, Papua New Guinea, Philippines, Samoa, São Tomé and Principe, Senegal, Solomon Islands, Sri Lanka, Sudan, Swaziland, Syrian Arab Republic, Tajikistan, Timor-Leste, Ukraine, Uzbekistan, Vanuatu, Vietnam, West Bank and Gaza, Yemen, Zambia. The category of *upper-middle-income economies* ($4,126 to $12,735) include the following countries: Albania, Algeria, American Samoa, Angola, Azerbaijan, Belarus, Belize, Bosnia and Herzegovina, Botswana, Brazil, Bulgaria, China, Colombia, Costa Rica, Cuba, Dominica, Dominican Republic, Ecuador, Fiji, Gabon, Grenada, Islamic Rep. of Iran, Iraq, Jamaica, Jordan, Kazakhstan, Lebanon, Libya, Macedonia, Malaysia, Maldives, Marshall Islands, Mauritius, Mexico, Mongolia, Montenegro, Namibia, Palau, Panama, Paraguay, Peru, Romania, Serbia, South Africa, St. Lucia, St. Vincent and the Grenadines, Suriname, Thailand, Tonga, Tunisia, Turkey, Turkmenistan, Tuvalu. The category of *high-income economies* ($12,736 or more) include the following countries: Andorra, Antigua and Barbuda, Argentina, Aruba, Australia, Austria, Bahamas, Bahrain, Barbados, Belgium, Bermuda, Brunei Darussalam, Canada, Cayman Islands, Channel Islands, Chile, Croatia, Curaçao, Cyprus, Czech Republic, Denmark, Estonia, Equatorial Guinea, Faeroe Islands, Finland, France, French Polynesia, Germany, Greece, Greenland, Guam, Hong Kong SAR China, Hungary, Iceland, Ireland, Isle of Man, Israel, Italy, Japan, Rep. Korea, Kuwait, Latvia, Liechtenstein, Lithuania, Luxembourg, Macao SAR China, Malta, Monaco, Netherlands, New Caledonia, New Zealand, Northern Mariana Islands, Norway, Oman, Poland, Portugal, Puerto Rico, Qatar, Russian Federation, San Marino, Saudi Arabia, Seychelles, Singapore, Sint Maarten (Dutch part), Slovak Republic, Slovenia, Spain, St. Kitts and Nevis, St. Martin (French part), Sweden, Switzerland, Taiwan-China, Trinidad and Tobago, Turks and Caicos Islands, United Arab Emirates, United Kingdom, United States, Uruguay, Venezuela RB, Virgin Islands (U.S.).

According to official statistics provided by World Bank for the current 2018 fiscal year, the following classification of countries applies, ie:

- *Low-income economies* are defined as those with a GNI per capita, calculated using the World Bank Atlas method, of $1,005 or less in 2016;
- *Lower middle-income economies* are defined as those with a GNI per capita, calculated using the World Bank Atlas method, between $1,006 and $3,955;
- *Upper middle-income economies* are defined as those with a GNI per capita, calculated using the World Bank Atlas method, between $3,956 and $12,235;
- *High-income economies* are defined as those with a GNI per capita, calculated using the World Bank Atlas method, of $12,236 or more.

Moreover, authors such as Hoskisson, Eden, Ming Lau and Wright (2000) suggested that "Emerging economies are low-income, rapid-growth countries using economic liberalization as their primary engine of growth." The authors also considered that emerging economies are divided into two major categories, namely: developing countries in Asia, Latin America, Africa, and the Middle East and transition economies in the former Soviet Union and China. The approach reflects the importance of emerging countries in the turbulent context of global economic changes.

O'Neill (2001) investigated the relationship between the G7 and some of the largest emerging market economies, ie Brazil, Russia, India and China (BRICs). Otherwise, O'Neill introduced a benchmark approach in literature using the acronym BRIC as a standard for emerging countries considering ongoing changes in the global economy. On the other hand, BRICS is an acronym that refers to certain countries such as Brazil, Russia, India, China and South Africa. The acronym BRICS it is actually an extension of the acronym BRIC by including another emerging countries ie South Africa on another continent such as Africa. The very concept of BRICS represents an alternative and a counterbalance to G8 and G20 which include some of the most developed countries in the world.

GLOBAL IMPLICATIONS OF SUSTAINABLE DEVELOPMENT

The United Nation established the Brundtland Commission in order to establish a union of countries aiming the sustainable development based on the report of the United Nations World Commission on Environment and Development entitled "Our Common Future" or the Brundtland Report (1987). Thus, the term sustainable development was highlighted with all its global implications. The global environmental challenges generated the necessity to adopt a new perspective based on sustainable development. Sustainable development, although it is a widely used concept, has many different meanings and therefore leads to many different answers. The concept of sustainability is based on three main pillars, ie: economic sustainability, social sustainability and ecological sustainability. Generalizing, the theoretical approaches of sustainability in order to identify the optimal tool to quantify the impact on natural environment.

Sustainable development has as its main objective to achieve, in a balanced manner, economic development, social development and environmental protection. A significant role in the theoretical outline of the concept of sustainability belongs to the term of economic sustainability. Synthesising, *economic sustainability* refers to the capacity or ability of an economy to support a certain level of economic production indefinitely in medium and long term. Implicitly, considering the economic and financial turmoil of recent years at the global level, a major importance targets economic sustainability based on certain positive implications for emerging or developing countries.

Social sustainability is the capacity of a social system, such as a country, a family or an organization, to function at a predetermined level of balance and social harmony for an indefinite period. On the other hand, an unsustainable social system is defined by negative aspects such as: war, endemic poverty, generalized social inequity, widespread injustice, precarious education and strong labor force migration. Therefore, it is relatively easy to identify which are the symptoms of an unsustainable social system, precisely because of it extremely dramatic impact and long-term effects. The concept of social sustainability is a theoretical conglomerate illustrating the diversity of nuances and implications which are essential for being properly understood. In this respect, the amplification of the phenomenon of poverty and social exclusion are decisive factors in the awareness of the importance of social sustainability.

Environmental sustainability is the ability of the natural environment to support a certain level of environmental quality and unrestrained natural resource extraction rates. Moreover, the intensification of severe environmental changes and the necessity of preserving existing natural resources constitute the major fundamental challenges in the current context of globalization.

According to World Bank's overview on sustainable development "growth must be both inclusive and environmentally sound to reduce poverty and build shared prosperity for today's population and to continue to meet the needs of future generations." Moreover, the general approach of World Bank highlights one of the fundamental traits of sustainable development, ie "it is efficient with resources and carefully planned to deliver both immediate and long-term benefits for people, planet, and prosperity."

The headline indicators for sustainable development include the following categories: real GDP per capita, resource productivity, people at risk of poverty or social exclusion, employment rate of older workers, life expectancy and healthy life years, greenhouse gas emissions, primary energy consumption, energy consumption of transport relative to GDP, common bird index, respectively official development assistance (European Union – EUROSTAT, 2015).

The long-term sustainability of infrastructure is extremely important in developing countries considering the growth potential and economic development opportunities. Moreover, sustainable development contributes to efficient use of resources as well as to optimize maintenance requirements. Sustainable development generates a great attention both in theory and practice. This laudatory reactions triggered by the fact that it has promoted economic growth, environmental well-being, while improving the welfare of citizens.

Furthermore, World Bank suggests that "the three pillars of sustainable development – economic growth, environmental stewardship, and social inclusion – carry across all sectors of development, from cities facing rapid urbanization to agriculture, infrastructure, energy development and use, water availability, and transportation."

E-SOCIETY: THE BORDER BETWEEN INNOVATIVE PANACEA AND NATURAL ALTERNATIVE TO CURRENT REALITY

E-Society is defined in very different manners so it is very difficult to provide an all-encompassing and generally accepted definition. The tangibility of E-Society is an abstract notion, but its importance is overwhelming for the emerging economies. E-Society is a community whose activity considerably involves digital transactions, computer networks, information and communication technology.

E-Society development led to the creation of new conceptual identity based on information and communication technologies. However, a significant dilemma highlights the need to prioritize both security and privacy. It should be noted that, in a heavily digitized world these issues become very vulnerable. A cybernetics approach captures the extreme implications regarding security and privacy issues in emerging economies. The importance of using internet or web-based networking infrastructure is rather significant especially in the case of emerging countries.

One of the essential characteristics of E-Society is constantly update knowledge based on virtual experience tools. The E-Society environment generates a virtual reality as effective alternative to traditional methods of approach. The importance of E-Society has increased considerably in recent years according to the heterogeneity of socio-economic environment and the challenges of globalization. Given the changing reality which is evolving extremely rapidly, technology should be considerably adapted. Technological tools are extremely important in sustaining the E-Society environment. An electronic society is not an entity in the real world, it is rather an abstract concept based on a virtual quintessence. In other words, virtual reality (VR) contributes significantly to the advancement of E-Society. In addition, the area of expertise of E-Society is very varied and include areas such as E-Tourism, E-Learning, E-Government, E-Health, E-Commerce, E-Education, E-Business.

DISSEMINATING INFRASTRUCTURE IMPLICATIONS

The building of infrastructure also require consistent mitigating expenditure. The significance of infrastructure is increasing in emerging countries given continued population growth and accelerated urbanization level. However, in most cases the existing infrastructure resources are not sufficient to achieve a satisfactory level of development. A characteristic of emerging countries is growing urbanization and the movement of population in urban rather than rural areas such as the need for basic infrastructure is more stringent. Practically are indispensable several aspects such as electricity utilities, natural gas, fresh water, sewage, transportation networks, telephone, television, internet connection and other basic amenities (see Figure 1).

Infrastructure helps determine the success of manufacturing and agricultural activities. Investments in water, sanitation, energy, housing, and transport also improve lives and help reduce poverty (World Bank, n.d). Moreover, internet users are individuals who have used the Internet (from any location) in the last 12 months. Internet can be used via a computer, mobile phone, personal digital assistant, games machine, digital TV etc. (World Bank, n.d).

Figure 2 servers per 1 million people, which are servers using encryption technology in Internet transactions, according to the definition provided by the World Bank official website.

Figure 3, which represent subscriptions to a public mobile telephone service that provide access to the PSTN using cellular technology, according to the definition provided by the World Bank official website.

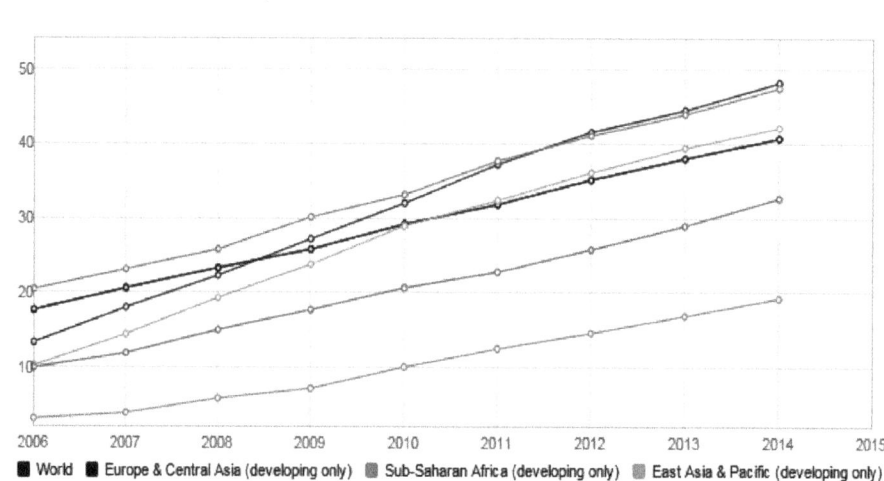

Figure 1. Comparative chart on Internet users (per 100 people)
Source: World Bank estimates, Data – Indicators

Figure 2. Secure Internet servers (per 1 million people)
Source: World Bank estimates, Data – Indicators

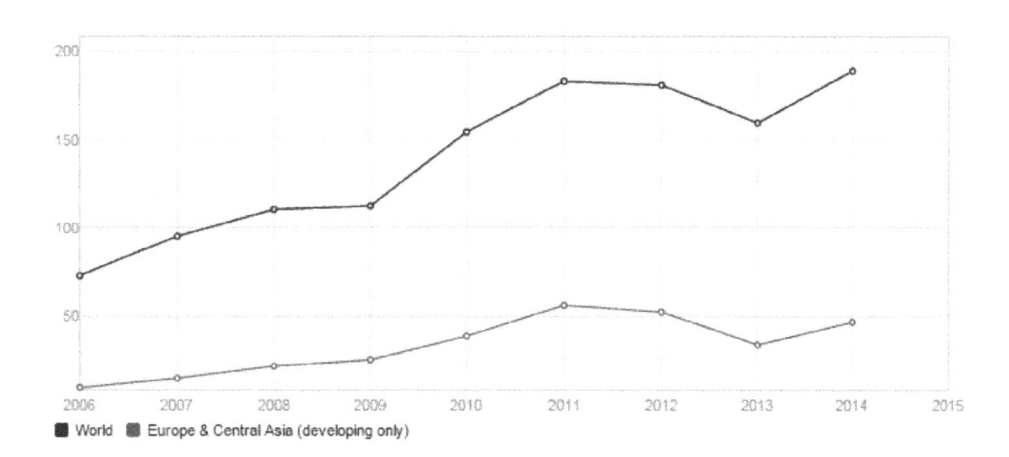

Figure 3. Mobile cellular subscriptions (per 100 people)
Source: World Bank estimates, Data – Indicators

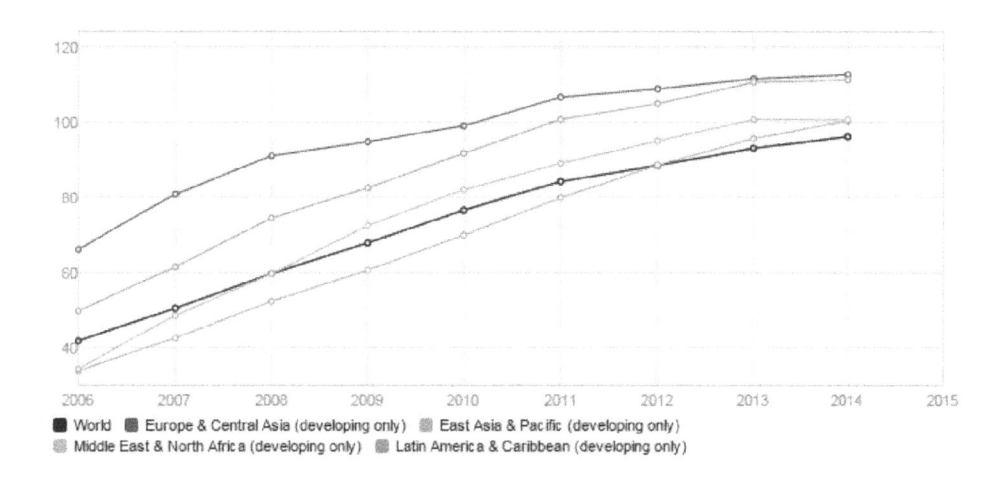

STRATEGIES AND GOALS ON SUSTAINABLE DEVELOPMENT

According to the official website of United Nations, on September 25th 2015, countries adopted a set of 17 goals to end poverty, protect the planet and ensure prosperity for each and everyone as part of a new sustainable development agenda, namely "Transforming our world: the 2030 Agenda for Sustainable Development". In other words, the 17 sustainable development goals represent actually identifying the key priorities for for humanity in the near future. The 17 goals set during this recent sustainable development agenda are the following:

- **Goal 1:** End poverty in all its forms everywhere
- **Goal 2:** End hunger, achieve food security and improved nutrition and promote sustainable agriculture
- **Goal 3:** Ensure healthy lives and promote well-being for all at all ages
- **Goal 4:** Ensure inclusive and quality education for all and promote lifelong learning
- **Goal 5:** Achieve gender equality and empower all women and girls
- **Goal 6:** Ensure access to water and sanitation for all
- **Goal 7:** Ensure access to affordable, reliable, sustainable and modern energy for all
- **Goal 8:** Promote inclusive and sustainable economic growth, employment and decent work for all
- **Goal 9:** Build resilient infrastructure, promote sustainable industrialization and foster innovation
- **Goal 10:** Reduce inequality within and among countries
- **Goal 11:** Make cities inclusive, safe, resilient and sustainable
- **Goal 12:** Ensure sustainable consumption and production patterns
- **Goal 13:** Take urgent action to combat climate change and its impacts
- **Goal 14:** Conserve and sustainably use the oceans, seas and marine resources
- **Goal 15:** Sustainably manage forests, combat desertification, halt and reverse land degradation, halt biodiversity loss
- **Goal 16:** Promote just, peaceful and inclusive societies
- **Goal 17:** Revitalize the global partnership for sustainable development

On the other hand, the World Bank Group has established a series of challenging but achievable targets regarding sustainable development. The World Bank Group includes 189 member countries, ie 188 United Nations countries and Kosovo) and aims to reduce poverty and support high rates of economic growth in developing, middle-income transition countries. The World Bank Group's mission statement released in 2013 provides a very ambitious perspective on effective implementation of sustainable development. Moreover, the World Bank Group integrates the principles of sustainable development mainly considering to end extreme poverty and promote shared prosperity. Accordingly, the World Bank Group seeks concrete goals developing main directions, ie the percentage of people living with less than $1.25 a day to fall to no more than 3 percent globally by 2030 (targeting the basic goal of ending extreme poverty) and achieving foster income growth of the bottom

40 percent of the population in every country (targeting the basic goal of promoting shared prosperity).

Statistically, the World Bank Group's mission statement released in 2013 highlights some very worrying percent regarding the evolution of extreme poverty in low income countries, most of which are part of Sub-Saharan Africa. Thereby, in 2010, more than one third of these states had a poverty rate above 50%. Even more dramatic is that 12 countries in Sub-Saharan Africa, the extreme poverty rate exceeds 60%, while four of these countries, ie Burundi, the Democratic Republic of Congo, Liberia, and Madagascar are in a state of extreme poverty which exceeds the 80% .

According to European Union – EUROSTAT 2015 monitoring report of the EU Sustainable Development Strategy, sustainable development indicators (SDI) include representative fields which is based on the economic, the social, the environmental, the global and the institutional areas, i.e.:

- Socio-economic development
- Sustainable consumption and production
- Social inclusion
- Demographic changes
- Public health
- Climate change and energy
- Sustainable transport
- Natural resources
- Global partnership
- Good governance.

The most recent strategy regarding sustainable development is focused on ending extreme poverty within a generation and promoting shared prosperity must be achieved in such a way as to be sustainable over time and across generations. This requires promoting environmental, social, and fiscal sustainability. We need to secure the long-term future of our planet and its resources so future generations do not find themselves in a wasteland. We also must aim for sustained social inclusion and limit the size of economic debt inherited by future generations. (World Bank, n.d)

Sustainable development allows for every human being to acquire the knowledge, skills, attitudes and values necessary to shape a sustainable future (UNESCO, n.d). In addition, the contribution of education for sustainable development targets "including key sustainable development issues into teaching and learning, for example, climate change, disaster risk reduction, biodiversity, poverty reduction, and sustainable consumption (UNESCO, n.d).

According to the outcome document of the United Nations Conference on Sustainable Development, entitled "The future we want" (Rio de Janeiro, 2012) common vision of the member states reveals that: "poverty eradication, changing unsustainable and promoting sustainable patterns of consumption and production and protecting and managing the natural resource base of economic and social development are the overarching objectives of and essential requirements for sustainable development." Moreover, the outcome document of the United Nations Conference on Sustainable Development, entitled "The future we want" (Rio de Janeiro, 2012) also highlighted that: "democracy, good governance and the rule of law, at the national and international levels, as well as an enabling environment, are essential for sustainable development, including sustained and inclusive economic growth, social development, environmental protection and the eradication of poverty and hunger".

CONCLUSION AND FUTURE DIRECTIONS

The global implications of sustainability and E-Society infrastructure in developing economies are extremely important. The preservation of non-renewable or finite natural resources and the health of the environment are fundamental directions in achieving sustainable development. Natural resources can also be renewable and naturally regenerated such as sunlight (solar energy), wind (wind energy, wind turbines), water (hydro energy), vegetable oil and animal fat (used to obtain biodiesel which can replace petroleum diesel or bioplastic) or many others. The conservation of exhaustible natural resources highlights a challenge of great current importance especially considering the intensification of the extreme weather phenomena in the last period. Enhancing the importance of social responsibility in order to optimize specific activity is a inherent consequence in the efficient implementation of public strategies for sustainable development.

An epistemological approach to the concept of economic sustainability involves a complex approach with significant implications in the medium and long term. Sustainable development and its complex socio-economic implications are above all a great challenge of the modern world, an altruistic alternative and a beneficial solution in numerous dramatic problems. Moreover, a paradox at the border between future and past, sustainable development is at the same time an awareness of the human impact on the natural environment. The global environmental challenges are very sharp in the case of emerging countries so that sustainable development plays an essential role. Future generations highly depend on the efficient implementation of measures on sustainable development. An E-Society is composed of e-communities based on information and communication technologies, also involving high connotations in the context of developing countries. Furthermore, using internet or web-based networking infrastructure represents a core feature in terms of e – society. An additional challenge requires a sustainable development in the case of extremely heterogeneous issues.

In the recent past, a great number of research studies have been published in the literature on the negative impact of poverty, the multidimensional effects of social exclusion and the major importance of sustainable development in the case of developing countries. Sustainable development contributes to improving the standard of living for the population in emerging countries without diminishing the chances of a similar existential condition for future generations. Moreover, economic growth and government policies regarding economic development in relation to the growth of an emerging country population imply significant changes in the social structure. Reducing or eradicating poverty has been a permanent desideratum of public management reform strategies in order to raise the living standards. An effective alternative to traditional policies and strategies is required in order to meet all the stringent needs of the current existing reality in most developing countries.

REFERENCES

Awasthi, P. (2011). Socio-Economic Challenges and Sustainable Development in Developing Countries. *Management Insight*, *7*(2), 56–63.

Bina, O. (2013). The green economy and sustainable development: An uneasy balance? *Environment and Planning. C, Government & Policy*, *31*(6), 1023–1047. doi:10.1068/c1310j

Birău, R. (2015). Statistical analysis on poverty in European Union. A case study for Romania. *International Journal of Business Quantitative Economics and Applied Management Research*, *1*(7), 1–10.

Birău, R. (2017) Conceptual approaches on sustainable development in public administration in Romania. *The Journal Contemporary Economy*, *2*(4), 34-37.

Bocchini, P., Frangopol, D., Ummenhofer, T., & Zinke, T. (2014). Resilience and Sustainability of Civil Infrastructure: Toward a Unified Approach. *Journal of Infrastructure Systems, Publisher: American Society of Civil Engineers*, *20*(2).

Castro, C. J. (2004). Sustainable Development: Mainstream and Critical Perspectives. *Organization & Environment*, *17*(2), 195–225. doi:10.1177/1086026604264910

Cheever, F., & Dernbach, J. C. (2015). Sustainable Development and its Discontents. *Journal of Transnational Environmental Law*.

Dale, A., & Newman, L. (2005). Sustainable development, education and literacy. *International Journal of Sustainability in Higher Education*, *6*(4), 351–362. doi:10.1108/14676370510623847

Dasgupta, S., & Tam, E. K. L. (2005). Indicators and framework for assessing sustainable infrastructure. *Canadian Journal of Civil Engineering*, *32*(1), 30–44. doi:10.1139/l04-101

El-Diraby, T., & Wang, B. (2005). E-Society Portal: Integrating Urban Highway Construction Projects into the Knowledge City. *Journal of Construction Engineering and Management*, *11*(1196), 1196-1211.

European Commission. (n.d.). *Eurostat*. Retrieved from http://ec.europa.eu/eurostat/

European Union – EUROSTAT. (2015). *Sustainable development in the European Union*. Luxembourg: Publications Office of the European Union. Retrieved from http://ec.europa.eu/eurostat/documents/3217494/6975281/KS-GT-15-001-EN-N.pdf

FTSE. (2018). *Country Classification*. Retrieved from http://www.ftse.com/products/indices/country-classification

George, C. (2007). Sustainable Development and Global Governance. *Journal of Environment & Development*, *16*(1), 102–125. doi:10.1177/1070496506298147

Hopwood, B., Mellor, M., & O'Brien, G. (2005). Sustainable development: Mapping different approaches. *Sustainable Development*, *13*(1), 38–52. doi:10.1002d.244

Hoskisson, R. E., Eden, L., Lau, C. M., & Wright, M. (2000). Strategy in Emerging Economies. *Academy of Management Journal*, *43*(3), 249–267. doi:10.2307/1556394

Howard, J., & Wheeler, J. (2015). What community development and citizen participation should contribute to the new global framework for sustainable development, *Community Development Journal. Oxford Journals*, *50*(4), 552–570.

Kaminsky, J. A., & Javernick-Will, A. N. (2014). The Internal Social Sustainability of Sanitation Infrastructure. *Environmental Science & Technology (ES&T)*, *48*(17), 10028–10035.

Leuenberger, D. Z., & Wakin, M. (2007). Sustainable Development in Public Administration Planning: An Exploration of Social Justice, Equity, and Citizen Inclusion. *Administrative Theory & Praxis Journal*, *29*(3), 394–411.

Longley, P. A., & Singleton, A. D. (2009). Classification through consultation: Public views of the geography of the E-Society. *International Journal of Geographical Information Science*, *23*(6), 737–763. doi:10.1080/13658810701704652

Magoulas, G., Lepouras, G., & Vassilakis, C. (2007). Virtual reality in the E-Society. *Virtual Reality Journal*, *11*(2), 71 – 73.

Mody, A. (2004). *What Is an Emerging Market?* IMF Working Paper, Research Department, WP/04/177, International Monetary Fund.

Momaya, K. (2011). Cooperation for competitiveness of emerging countries: Learning from a case of nanotechnology. *Competitiveness Review*, *21*(2), 152–170. doi:10.1108/10595421111117443

Newman, P. (2016). Sustainable Urbanization: Four stages of Infrastructure Planning and Progress. *The Journal of Sustainable Urbanization Planning and Progress*, *1*(1), 3–10.

Newman, P. W. (2015). Transport infrastructure and sustainability: A new planning and assessment framework. *Smart and Sustainable Built Environment*, *4*(2), 140–153.

O'Neill, J. (2001). *Building Better Global Economic BRICs.* Global Economics, Paper No: 66, Goldman Sachs. Retrieved from https://www.gs.com

Ogbodo, S. G. (2010). The Paradox of the Concept of Sustainable Development under Nigeria's Environmental Law. *Journal of Sustainable Development*, *3*(3). doi:10.5539/jsd.v3n3p201

Otegbulu, A., & Adewunmi, Y. (2009). Evaluating the sustainability of urban housing development in Nigeria through innovative infrastructure management. *International Journal of Housing Markets and Analysis*, *2*(4), 334–346. doi:10.1108/17538270910992782

Pour-Ghaz, M. (2013). Sustainable Infrastructure Materials: Challenges and Opportunities. *International Journal of Applied Ceramic Technology*, *10*(4), 584–592. doi:10.1111/ijac.12083

Raiyn, J. (2011). Developing E-Society Cognitive Platform Based on the Social Agent E-learning Goal Oriented. *Advances in Internet of Things*, *1*(1), 1–4. doi:10.4236/ait.2011.11001

Rosa, D. J. (2009). Sustainability and Infrastructure Resource Allocation. *Journal of Business & Economics Research*, *7*(9).

Schmid, B., Stanoevska, K., & Tschammer, V. (2013). *Towards the E-Society: E-Commerce, E-Business, and E-Government.* Springer Publishing Company, Incorporated.

Shen, L., Wu, Y., & Zhang, X. (2011). Key Assessment Indicators for the Sustainability of Infrastructure Projects. *Journal of Construction Engineering and Management, Publisher American Society of Civil Engineers, 137*(6).

Siew, R. Y. J., Balatbat, M. C. A., & Carmichael, D. G. (2016). A proposed framework for assessing the sustainability of infrastructure. *International Journal of Construction Management*.

Simon, Y. (1997). Encyclopédie des marchés financiers (Vols. 1-2). Editeur Economica.

Sustainable Development Knowledge Platform, Division For Sustainable Development, UN-DESA. (n.d.). The United Nations, Department of Economic and Social Affairs. Retrieved from https://sustainabledevelopment.un.org/

Swilling, M. (2006). Sustainability and infrastructure planning in South Africa: A Cape Town case study. *Environment and Urbanization, 18*(1), 23–50. doi:10.1177/0956247806063939

Tan, S. (2014). *Challenges and Issues of Sustainable Development Policies.* Lee Kuan Yew School of Public Policy Research Paper No. 14-25. Available at SSRN: http://ssrn.com/abstract=2484930

Ugwu, O. O. (2013). Nanotechnology for sustainable infrastructure in 21st century civil engineering. *Journal of Innovative Engineering, 1*(1).

UNESCO. (n.d.). *UNESCO*. Retrieved from www.unesco.org

United Nations. (n.d.). *United Nations*. Retrieved from http://www.un.org/

Wehrli, A. (2014). Why Mountains Matter for Sustainable Development. *Mountain Research and Development, 34*(4), 405–409. doi:10.1659/MRD-JOURNAL-D-14-00096.1

Wessels, M. (2014) Stimulating sustainable infrastructure development through public–private partnerships. *Proceedings of the Institution of Civil Engineers - Management, Procurement and Law, 167*(5), 232 – 241.

World Bank. (n.d.). *World Bank Open Data*. Retrieved from http://data.worldbank.org/

Zheng, Y., & Walsham, G. (2008). Inequality of what? Social exclusion in the E-Society as capability deprivation. *Information Technology & People, 21*(3), 222–243. doi:10.1108/09593840810896000

KEY TERMS AND DEFINITIONS

Developing Economy: Describes the economy of an emerging country and it is characterized by a lower standard of living, a lower human development index (HDI) and a lower quality of life.

E-Society Infrastructure: Based on a number of communication and technical devices in order to develop a sustainable environment, such as: mobile phone devices such as iPads, iPods, handheld tablet, notebook, mobile phones, mobile PC, smartphone and other portable accessories, or essential components such as: wireless technology, social media technology (networking), satellite digital services, computer systems technology, multimedia technologies, information technology, and others.

Emerging Countries: A category of country classification characterized by a lower standard of living and quality of life, an increased level of poverty, increased financial volatility, significant social inequity, a low level of industrialization, a fragile infrastructure, modest architecture of the health system, high fragmentation, a higher birth rate, inadequate education systems, a higher infant mortality rate.

Globalization: The modern term used to describe changes in the structure of societies and the world economy, but having a major impact in the context of an accelerated informatization.

Sustainability: Intended to provide an optimal solution to negative phenomena such as massive exploitation of natural resources and the continuous degradation of natural environment.

Sustainable Development: Based on three main pillars (i.e., economic sustainability, social sustainability, and ecological sustainability).

Related References

To continue our tradition of advancing research in the area of environmental science and technologies, we have compiled a list of recommended IGI Global readings. These references will provide additional information and guidance to further enrich your knowledge and assist you with your own research and future publications.

Abayomi, K., de la Pena, V., Lall, U., & Levy, M. (2011). Quantifying sustainability: Methodology for and determinants of an environmental sustainability index. In Z. Luo (Ed.), *Green finance and sustainability: Environmentally-aware business models and technologies* (pp. 74–89). Hershey, PA: IGI Global. doi:10.4018/978-1-60960-531-5.ch004

Abdel Gelil, I. (2012). Globalization of the environmental issues: Response of the Arab region. In M. Tortora (Ed.), *Sustainable systems and energy management at the regional level: Comparative approaches* (pp. 147–165). Hershey, PA: IGI Global. doi:10.4018/978-1-61350-344-7.ch008

Adewumi, J., Ilemobade, A., & van Zyl, J. (2013). Application of a multi-criteria decision support tool in assessing the feasibility of implementing treated wastewater reuse. *International Journal of Decision Support System Technology*, 5(1), 1–23. doi:10.4018/jdsst.2013010101

Adeyemo, J., Adeyemo, F., & Otieno, F. (2012). Assessment of pollutant loads of runoff in Pretoria, South Africa. In E. Carayannis (Ed.), *Sustainable policy applications for social ecology and development* (pp. 115–127). Hershey, PA: IGI Global. doi:10.4018/978-1-4666-1586-1.ch009

Ahmad, A. F., & Panni, M. F. (2014). Green marketing strategy: A pedagogical view. In H. Kaufmann & M. Panni (Eds.), *Handbook of research on consumerism in business and marketing: Concepts and practices* (pp. 92–124). Hershey, PA: IGI Global. doi:10.4018/978-1-4666-5880-6.ch005

Ahmad, M. F., & Siang, A. Y. (2014). Modelling of hydrodynamics and sediment transport at Pantai Tok Jembal, Kuala Terengganu Mengabang Telipot, Terengganu, using MIKE 21. In O. Olanrewaju, A. Saharuddin, A. Ab Kader, & W. Wan Nik (Eds.), *Marine technology and sustainable development: Green innovations* (pp. 109–126). Hershey, PA: IGI Global. doi:10.4018/978-1-4666-4317-8.ch007

Ahmed Al-kerdawy, M. M. (2011). The role of environmental innovation strategy in reinforcing the impact of green managerial practices on competitive advantages of fertilizer companies in Egypt. *International Journal of Customer Relationship Marketing and Management*, *2*(1), 36–54. doi:10.4018/jcrmm.2011010103

Akkaya, C., Wolf, P., & Krcmar, H. (2011). Efficient information provision for environmental and sustainability reporting. In I. Management Association (Ed.), Green technologies: Concepts, methodologies, tools and applications (pp. 1587-1609). Hershey, PA: IGI Global. doi:10.4018/978-1-60960-472-1.ch705

Al-kerdawy, M. M. (2013). The role of environmental innovation strategy in reinforcing the impact of green managerial practices on competitive advantages of fertilizer companies in Egypt. In R. Eid (Ed.), *Managing customer trust, satisfaction, and loyalty through information communication technologies* (pp. 37–53). Hershey, PA: IGI Global. doi:10.4018/978-1-4666-3631-6.ch003

Alves de Lima, A., Carvalho dos Reis, P., Branco, J. C., Danieli, R., Osawa, C. C., Winter, E., & Santos, D. A. (2013). Scenario-patent protection compared to climate change: The case of green patents. *International Journal of Social Ecology and Sustainable Development*, *4*(3), 61–70. doi:10.4018/jsesd.2013070105

Ang, Z. (2011). The impact of electricity market and environmental regulation on carbon capture & storage (CCS) development in China. In Z. Luo (Ed.), *Green finance and sustainability: Environmentally-aware business models and technologies* (pp. 463–471). Hershey, PA: IGI Global. doi:10.4018/978-1-60960-531-5.ch024

Antonova, A. (2013). Green, sustainable, or clean: What type of IT/IS technologies will we need in the future? In P. Ordóñez de Pablos (Ed.), *Green technologies and business practices: An IT approach* (pp. 151–162). Hershey, PA: IGI Global. doi:10.4018/978-1-4666-1972-2.ch008

Antonova, A. (2014). Green, sustainable, or clean: What type of IT/IS technologies will we need in the future? In *Sustainable practices: Concepts, methodologies, tools and applications* (pp. 384–396). Hershey, PA: IGI Global. doi:10.4018/978-1-4666-4852-4.ch021

Appiah, D. O., & Kemausuor, F. (2012). Energy, environment and socio-economic development: Africa's triple challenge and options. In M. Tortora (Ed.), *Sustainable systems and energy management at the regional level: Comparative approaches* (pp. 166–182). Hershey, PA: IGI Global. doi:10.4018/978-1-61350-344-7.ch009

Appleby, M. R., Lambert, C. G., Rennie, A. E., & Buckley, A. B. (2011). An investigation into the environmental impact of product recovery methods to support sustainable manufacturing within small and medium-sized enterprises (SMEs). *International Journal of Manufacturing, Materials, and Mechanical Engineering*, *1*(2), 1–18. doi:10.4018/ijmmme.2011040101

Appleby, M. R., Lambert, C. G., Rennie, A. E., & Buckley, A. B. (2013). An investigation into the environmental impact of product recovery methods to support sustainable manufacturing within small and medium-sized enterprises (SMEs). In J. Davim (Ed.), *Dynamic methods and process advancements in mechanical, manufacturing, and materials engineering* (pp. 73–90). Hershey, PA: IGI Global. doi:10.4018/978-1-4666-1867-1.ch004

Arora, K., Kumar, A., & Sharma, S. (2012). Energy from waste: Present scenario, challenges, and future prospects towards sustainable development. In P. Olla (Ed.), *Global sustainable development and renewable energy systems* (pp. 271–296). Hershey, PA: IGI Global. doi:10.4018/978-1-4666-1625-7.ch014

Arora, K., Kumar, A., & Sharma, S. (2014). Energy from waste: present scenario, challenges, and future prospects towards sustainable development. In *Sustainable practices: Concepts, methodologies, tools and applications* (pp. 1519–1543). Hershey, PA: IGI Global. doi:10.4018/978-1-4666-4852-4.ch085

Ashraf, G. Y. (2012). A study of eco-friendly supply chain management at cement industries of Chhattisgarh. In M. Garg & S. Gupta (Eds.), *Cases on supply chain and distribution management: Issues and principles* (pp. 146–157). Hershey, PA: IGI Global. doi:10.4018/978-1-4666-0065-2.ch007

Ashraf, G. Y. (2013). A study of eco-friendly supply chain management at cement industries of Chhattisgarh. In *Supply chain management: Concepts, methodologies, tools, and applications* (pp. 823–830). Hershey, PA: IGI Global. doi:10.4018/978-1-4666-2625-6.ch048

Asiimwe, E. N., & Åke, G. (2012). E-waste management in East African community. In K. Bwalya & S. Zulu (Eds.), *Handbook of research on e-government in emerging economies: Adoption, e-participation, and legal frameworks* (pp. 307–327). Hershey, PA: IGI Global. doi:10.4018/978-1-4666-0324-0.ch015

Aspradaki, A. A. (2013). Deliberative democracy and nanotechnologies in health. *International Journal of Technoethics*, *4*(2), 1–14. doi:10.4018/jte.2013070101

Ayadi, F. S. (2013). An empirical investigation of environmental kuznets curve in Nigeria. In K. Ganesh & S. Anbuudayasankar (Eds.), *International and interdisciplinary studies in green computing* (pp. 302–310). Hershey, PA: IGI Global. doi:10.4018/978-1-4666-2646-1.ch022

Ayuk, E. T., Fonta, W. M., & Kouame, E. B. (2014). Application of quantitative methods in natural resource management in Africa: A review. In P. Schaeffer & E. Kouassi (Eds.), *Econometric methods for analyzing economic development* (pp. 205–234). Hershey, PA: IGI Global. doi:10.4018/978-1-4666-4329-1.ch013

Bachour, N. (2012). Green IT project management: Optimizing the value of green IT projects within organizations. In W. Hu & N. Kaabouch (Eds.), *Sustainable ICTs and management systems for green computing* (pp. 146–178). Hershey, PA: IGI Global. doi:10.4018/978-1-4666-1839-8.ch007

Baginetas, K. N. (2011). Sustainable management of agricultural resources and the need for stakeholder participation for the developing of appropriate sustainability indicators: The case of soil quality. In Z. Andreopoulou, B. Manos, N. Polman, & D. Viaggi (Eds.), *Agricultural and environmental informatics, governance and management: Emerging research applications* (pp. 227–261). Hershey, PA: IGI Global. doi:10.4018/978-1-60960-621-3.ch013

Baginetas, K. N. (2012). Sustainable management of agricultural resources and the need for stakeholder participation for the developing of appropriate sustainability indicators: The case of soil quality. In *Regional development: Concepts, methodologies, tools, and applications* (pp. 632–665). Hershey, PA: IGI Global. doi:10.4018/978-1-4666-0882-5.ch401

Bailis, R., & Arabshahi, N. (2011). Voluntary emissions reduction: Are we making progress? In Z. Luo (Ed.), *Green finance and sustainability: Environmentally-aware business models and technologies* (pp. 241–273). Hershey, PA: IGI Global. doi:10.4018/978-1-60960-531-5.ch014

Banerjee, S., Sing, T. Y., Chowdhury, A. R., & Anwar, H. (2013). Motivations to adopt green ICT: A tale of two organizations. *International Journal of Green Computing*, *4*(2), 1–11. doi:10.4018/jgc.2013070101

Baptiste, A. K. (2013). Local vs. expert perception of climate change: An analysis of fishers in Trinidad and Tobago. In H. Muga & K. Thomas (Eds.), *Cases on the diffusion and adoption of sustainable development practices* (pp. 44–82). Hershey, PA: IGI Global. doi:10.4018/978-1-4666-2842-7.ch003

Bassey, K., & Chigbu, P. (2013). Optimal detection and estimation of marine oil spills through coherent pluralism. *International Journal of Operations Research and Information Systems*, *4*(1), 84–111. doi:10.4018/joris.2013010105

Bata, R., & Jordão, T. C. (2011). Modeling the influences of heating fuel consumption in gaseous emissions and solid waste generation. In V. Olej, I. Obršálová, & J. Krupka (Eds.), *Environmental modeling for sustainable regional development: System approaches and advanced methods* (pp. 162–185). Hershey, PA: IGI Global. doi:10.4018/978-1-60960-156-0.ch008

Beall, A. M., & Ford, A. (2012). Reports from the field: Assessing the art and science of participatory environmental modeling. In J. Wang (Ed.), *Societal impacts on information systems development and applications* (pp. 195–213). Hershey, PA: IGI Global. doi:10.4018/978-1-4666-0927-3.ch013

Ben Brahim, H., & Duckstein, L. (2011). Descriptive methods and compromise programming for promoting agricultural reuse of treated wastewater. In H. do Prado, A. Barreto Luiz, & H. Filho (Eds.), *Computational methods for agricultural research: Advances and applications* (pp. 355–388). Hershey, PA: IGI Global. doi:10.4018/978-1-61692-871-1.ch017

Bentley, G. C., Cromley, R. G., Hanink, D. M., & Heidkamp, C. P. (2013). Forest cover change in the northeastern U.S.: A spatial assessment in the context of an environmental kuznets curve. *International Journal of Applied Geospatial Research*, *4*(3), 1–18. doi:10.4018/jagr.2013070101

Berke, M. Ö., Sütlü, E., Avcioglu, B., & Gem, E. (2013). Identification of priority areas for conservation in Lake Egirdir and Lake Kovada, Turkey. In J. Papathanasiou, B. Manos, S. Arampatzis, & R. Kenward (Eds.), *Transactional environmental support system design: Global solutions* (pp. 199–202). Hershey, PA: IGI Global. doi:10.4018/978-1-4666-2824-3.ch018

Bhattarai, N., Khanal, S., Pudasaini, P. R., Pahl, S., & Romero-Urbina, D. (2011). Citrate stabilized silver nanoparticles: Study of crystallography and surface properties. *International Journal of Nanotechnology and Molecular Computation*, *3*(3), 15–28. doi:10.4018/ijnmc.2011070102

Bier, A. (2012). A system dynamics approach to changing perceptions about thermal water quality trading markets. In J. Wang (Ed.), *Societal impacts on information systems development and applications* (pp. 182–194). Hershey, PA: IGI Global. doi:10.4018/978-1-4666-0927-3.ch012

Biller, D., & Sanchez-Triana, E. (2013). Enlisting markets in the conservation and sustainable use of biodiversity in South Asia's Sundarbans. *International Journal of Social Ecology and Sustainable Development, 4*(3), 71–86. doi:10.4018/jsesd.2013070106

Bonadiman, R. (2013). Sustainability: Brazilian perspectives and challenges after the first kioto's protocol period. *International Journal of Social Ecology and Sustainable Development, 4*(3), 52–60. doi:10.4018/jsesd.2013070104

Boote, K. J., Jones, J. W., Hoogenboom, G., & White, J. W. (2012). The role of crop systems simulation in agriculture and environment. In P. Papajorgji & F. Pinet (Eds.), *New technologies for constructing complex agricultural and environmental systems* (pp. 326–339). Hershey, PA: IGI Global. doi:10.4018/978-1-4666-0333-2.ch018

Boulton, A., Devriendt, L., Brunn, S. D., Derudder, B., & Witlox, F. (2014). City networks in cyberspace and time: Using Google hyperlinks to measure global economic and environmental crises. In *Crisis management: Concepts, methodologies, tools and applications* (pp. 1325–1345). Hershey, PA: IGI Global. doi:10.4018/978-1-4666-4707-7.ch067

Bradbury, M. (2012). The sustainable waterfront. In O. Ercoskun (Ed.), *Green and ecological technologies for urban planning: Creating smart cities* (pp. 274–292). Hershey, PA: IGI Global. doi:10.4018/978-1-61350-453-6.ch015

Bradbury, M. (2014). The sustainable waterfront. In *Sustainable practices: Concepts, methodologies, tools and applications* (pp. 1683–1700). Hershey, PA: IGI Global. doi:10.4018/978-1-4666-4852-4.ch093

Brimmo, A., & Emziane, M. (2014). Carbon nanotubes for photovoltaics. In M. Bououdina & J. Davim (Eds.), *Handbook of research on nanoscience, nanotechnology, and advanced materials* (pp. 268–311). Hershey, PA: IGI Global. doi:10.4018/978-1-4666-5824-0.ch012

Brister, E., Hane, E., & Korfmacher, K. (2011). Visualizing plant community change using historical records. *International Journal of Applied Geospatial Research, 2*(4), 1–18. doi:10.4018/jagr.2011100101

Buxton, G. (2013). Nanotechnology and polymer solar cells. In S. Anwar, H. Efstathiadis, & S. Qazi (Eds.), *Handbook of research on solar energy systems and technologies* (pp. 231–253). Hershey, PA: IGI Global. doi:10.4018/978-1-4666-1996-8.ch009

Buxton, G. (2014). Nanotechnology and polymer solar cells. In Nanotechnology: Concepts, methodologies, tools, and applications (pp. 384-405). Hershey, PA: IGI Global. doi:10.4018/978-1-4666-5125-8.ch015

Cadman, T., & Hume, M. (2012). Developing sustainable governance systems for regional sustainability programmes and 'green' business practices: The case of 'green' timber. In M. Tortora (Ed.), *Sustainable systems and energy management at the regional level: Comparative approaches* (pp. 365–382). Hershey, PA: IGI Global. doi:10.4018/978-1-61350-344-7.ch019

Cai, T. (2014). Artificial neural network for industrial and environmental research via air quality monitoring network. In Z. Sun & J. Yearwood (Eds.), *Handbook of research on demand-driven web services: Theory, technologies, and applications* (pp. 399–419). Hershey, PA: IGI Global. doi:10.4018/978-1-4666-5884-4.ch019

Cai, T. (2014). Geospatial technology-based e-government design for environmental protection and emergency response. In K. Bwalya (Ed.), *Technology development and platform enhancements for successful global e-government design* (pp. 157–184). Hershey, PA: IGI Global. doi:10.4018/978-1-4666-4900-2.ch009

Calipinar, H., & Ulas, D. (2013). Model suggestion for SMEs economic and environmental sustainable development. In N. Ndubisi & S. Nwankwo (Eds.), *Enterprise development in SMEs and entrepreneurial firms: Dynamic processes* (pp. 270–290). Hershey, PA: IGI Global. doi:10.4018/978-1-4666-2952-3.ch014

Carvajal-Escobar, Y., Mimi, Z., Khayat, S., Sulieman, S., Garces, W., & Cespedes, G. (2011). Application of methodologies for environmental flow determination in an andean and a Mediterranean basin: Two case studies of the Pance River (Colombia) and Wadi River (Palestine) basin. *International Journal of Social Ecology and Sustainable Development*, 2(4), 26–43. doi:10.4018/jsesd.2011100103

Carvajal-Escobar, Y., Mimi, Z., Khayat, S., Sulieman, S., Garces, W., & Cespedes, G. (2013). Application of methodologies for environmental flow determination in an andean and a Mediterranean basin: Two case studies of the Pance River (Colombia) and Wadi River (Palestine) basin. In E. Carayannis (Ed.), *Creating a sustainable ecology using technology-driven solutions* (pp. 296–314). Hershey, PA: IGI Global. doi:10.4018/978-1-4666-3613-2.ch020

Cascelli, E., Crestaz, E., & Tatangelo, F. (2013). Cartography and geovisualization in groundwater modelling. In G. Borruso, S. Bertazzon, A. Favretto, B. Murgante, & C. Torre (Eds.), *Geographic information analysis for sustainable development and economic planning: New technologies* (pp. 49–67). Hershey, PA: IGI Global. doi:10.4018/978-1-4666-1924-1.ch004

Ceccaroni, L., & Oliva, L. (2012). Ontologies for the design of ecosystems. In T. Podobnikar & M. Čeh (Eds.), *Universal ontology of geographic space: semantic enrichment for spatial data* (pp. 207–228). Hershey, PA: IGI Global. doi:10.4018/978-1-4666-0327-1.ch009

Charuvilayil, R. A. (2013). Industrial pollution and people's movement: A case study of Eloor Island Kerala, India. In H. Muga & K. Thomas (Eds.), *Cases on the diffusion and adoption of sustainable development practices* (pp. 312–351). Hershey, PA: IGI Global. doi:10.4018/978-1-4666-2842-7.ch012

Chen, E. T. (2011). Green information technology and virtualization in corporate environmental management information systems. In *Green technologies: Concepts, methodologies, tools and applications* (pp. 1421–1434). Hershey, PA: IGI Global. doi:10.4018/978-1-60960-472-1.ch605

Chen, H., & Bishop, I. D. (2013). Collaborative environmental knowledge management. *International Journal of E-Planning Research*, *2*(1), 58–81. doi:10.4018/ijepr.2013010104

Chen, X. M. (2011). GIS and remote sensing in environmental risk assessment. In *Green technologies: Concepts, methodologies, tools and applications* (pp. 840–847). Hershey, PA: IGI Global. doi:10.4018/978-1-60960-472-1.ch415

Chen, Y. (2013). Generalize key requirements for designing IT-based system for green with considering stakeholder needs. *International Journal of Information Technologies and Systems Approach*, *6*(1), 78–97. doi:10.4018/jitsa.2013010105

Chinchuluun, A., Xanthopoulos, P., Tomaino, V., & Pardalos, P. (2012). Data mining techniques in agricultural and environmental sciences. In P. Papajorgji & F. Pinet (Eds.), *New technologies for constructing complex agricultural and environmental systems* (pp. 311–325). Hershey, PA: IGI Global. doi:10.4018/978-1-4666-0333-2.ch017

Chitra, S. (2011). Adopting green ICT in business. In I. Management Association (Ed.), Green technologies: Concepts, methodologies, tools and applications (pp. 1145-1153). Hershey, PA: IGI Global. doi:10.4018/978-1-60960-472-1.ch501

Chiu, M. (2013). Gaps between valuing and purchasing green-technology products: Product and gender differences. *International Journal of Technology and Human Interaction*, *8*(3), 54–68. doi:10.4018/jthi.2012070106

Cho, C. H., Patten, D. M., & Roberts, R. W. (2014). Environmental disclosures and impression management. In R. Hart (Ed.), *Communication and language analysis in the corporate world* (pp. 217–231). Hershey, PA: IGI Global. doi:10.4018/978-1-4666-4999-6.ch013

Christophoridis, C., Bizani, E., & Fytianos, K. (2011). Environmental quality monitoring, using GIS as a tool of visualization, management and decision-making: Applications emerging from the EU water framework directive EU 2000/60. In Z. Andreopoulou, B. Manos, N. Polman, & D. Viaggi (Eds.), *Agricultural and environmental informatics, governance and management: Emerging research applications* (pp. 397–424). Hershey, PA: IGI Global. doi:10.4018/978-1-60960-621-3.ch021

Christophoridis, C., Bizani, E., & Fytianos, K. (2013). Environmental quality monitoring, using GIS as a tool of visualization, management and decision-making: Applications emerging from the EU water framework directive EU 2000/60. In *Geographic information systems: Concepts, methodologies, tools, and applications* (pp. 1559–1586). Hershey, PA: IGI Global. doi:10.4018/978-1-4666-2038-4.ch094

Cincu, C., & Diacon, A. (2013). Hybrid solar cells: Materials and technology. In L. Fara & M. Yamaguchi (Eds.), *Advanced solar cell materials, technology, modeling, and simulation* (pp. 79–100). Hershey, PA: IGI Global. doi:10.4018/978-1-4666-1927-2.ch006

Cocozza, A., & Ficarella, A. (2013). Electrical resistivity measures in cohesive soils for the simulation of an integrated energy system between CCS and low-enthalpy geothermal. *International Journal of Measurement Technologies and Instrumentation Engineering*, *3*(1), 48–68. doi:10.4018/ijmtie.2013010105

Cohen, E., & Zimmerman, T. D. (2012). Teaching the greenhouse effect with inquiry-based computer simulations: A WISE case study. In L. Lennex & K. Nettleton (Eds.), *Cases on inquiry through instructional technology in math and science* (pp. 551–580). Hershey, PA: IGI Global. doi:10.4018/978-1-4666-0068-3.ch020

Congedo, L., Baiocco, F., Brini, S., Liberti, L., & Munafò, M. (2013). Urban environment quality in the Italian spatial data infrastructure. In G. Borruso, S. Bertazzon, A. Favretto, B. Murgante, & C. Torre (Eds.), *Geographic information analysis for sustainable development and economic planning: New technologies* (pp. 179–192). Hershey, PA: IGI Global. doi:10.4018/978-1-4666-1924-1.ch012

Cosmi, C., Di Leo, S., Loperte, S., Pietrapertosa, F., Salvia, M., Macchiato, M., & Cuomo, V. (2011). Comprehensive energy systems analysis support tools for decision making. In *Green technologies: Concepts, methodologies, tools and applications* (pp. 493–514). Hershey, PA: IGI Global. doi:10.4018/978-1-60960-472-1.ch307

Cotton, M. (2012). Community opposition and public engagement with wind energy in the UK. In M. Tortora (Ed.), *Sustainable systems and energy management at the regional level: Comparative approaches* (pp. 310–327). Hershey, PA: IGI Global. doi:10.4018/978-1-61350-344-7.ch016

Cuccurullo, S., Francese, R., Passero, I., & Tortora, G. (2013). A 3D serious city building game on waste disposal. *International Journal of Distance Education Technologies*, *11*(4), 112–135. doi:10.4018/ijdet.2013100108

Cunha, M. D. (2011). Wastewater systems management at the regional level. In V. Olej, I. Obršálová, & J. Krupka (Eds.), *Environmental modeling for sustainable regional development: System approaches and advanced methods* (pp. 186–203). Hershey, PA: IGI Global. doi:10.4018/978-1-60960-156-0.ch009

Cunha, M. D. (2012). Wastewater systems management at the regional level. In *Regional development: Concepts, methodologies, tools, and applications* (pp. 1161–1177). Hershey, PA: IGI Global. doi:10.4018/978-1-4666-0882-5.ch607

Da Ronch, B., Di Maria, E., & Micelli, S. (2013). Clusters go green: Drivers of environmental sustainability in local networks of SMEs. *International Journal of Information Systems and Social Change*, *4*(1), 37–52. doi:10.4018/jissc.2013010103

Danahy, J., Wright, R., Mitchell, J., & Feick, R. (2013). Exploring ways to use 3D urban models to visualize multi-scalar climate change data and mitigation change models for e-planning. *International Journal of E-Planning Research*, *2*(2), 1–17. doi:10.4018/ijepr.2013040101

Dhal, S. (2013). Indigenous agricultural knowledge and innovation: A study of agricultural scientists in odisha. *International Journal of Information Systems and Social Change*, *4*(3), 57–71. doi:10.4018/jissc.2013070104

Dimitriou, D., Voskaki, A., & Sartzetaki, M. (2014). Airports environmental management: Results from the evaluation of European airports environmental plans. *International Journal of Information Systems and Supply Chain Management*, *7*(1), 1–14. doi:10.4018/IJISSCM.2014010101

Dizdaroglu, D., Yigitcanlar, T., & Dawes, L. (2011). Planning for sustainable urban futures. In *Green technologies: Concepts, methodologies, tools and applications* (pp. 1922–1932). Hershey, PA: IGI Global. doi:10.4018/978-1-60960-472-1.ch806

Djeflat, A. (2014). Harnessing knowledge for sustainable development: Challenges and opportunities for Arab countries. In A. Driouchi (Ed.), *Knowledge-based economic policy development in the Arab world* (pp. 229–244). Hershey, PA: IGI Global. doi:10.4018/978-1-4666-5210-1.ch009

Dolney, T. J. (2011). A GIS methodology for assessing the safety hazards of abandoned mine lands (AMLs): Application to the state of Pennsylvania. *International Journal of Applied Geospatial Research, 2*(3), 50–71. doi:10.4018/jagr.2011070104

Dubovski, S. (2014). Activities in oil and gas processing for avoiding or minimizing environmental impacts. In D. Matanovic, N. Gaurina-Medjimurec, & K. Simon (Eds.), *Risk analysis for prevention of hazardous situations in petroleum and natural gas engineering* (pp. 247–263). Hershey, PA: IGI Global. doi:10.4018/978-1-4666-4777-0.ch012

Dugas, D. P., DeMers, M. N., Greenlee, J. C., Whitford, W. G., & Klimaszewski-Patterson, A. (2011). Rapid evaluation of arid lands (REAL): A methodology. *International Journal of Applied Geospatial Research, 2*(3), 32–49. doi:10.4018/jagr.2011070103

Dusmanescu, D. (2013). Aspects regarding implementation of renewable energy sources in Romania up to 2050. *International Journal of Sustainable Economies Management, 2*(4), 1–21. doi:10.4018/ijsem.2013100101

Ehlinger, T., Tofan, L., Bucur, M., Enz, J., Carlson, J., & Shaker, R. (2011). Application of a participatory ex ante assessment model for environmental governance and visualizing sustainable redevelopment in Gorj County, Romania. In Z. Andreopoulou, B. Manos, N. Polman, & D. Viaggi (Eds.), *Agricultural and environmental informatics, governance and management: Emerging research applications* (pp. 61–86). Hershey, PA: IGI Global. doi:10.4018/978-1-60960-621-3.ch004

Ehlinger, T., Tofan, L., Bucur, M., Enz, J., Carlson, J., & Shaker, R. (2012). Application of a participatory ex ante assessment model for environmental governance and visualizing sustainable redevelopment in Gorj County, Romania. In *Regional development: Concepts, methodologies, tools, and applications* (pp. 743–768). Hershey, PA: IGI Global. doi:10.4018/978-1-4666-0882-5.ch407

Ekekwe, N. (2013). Nanotechnology and microelectronics: The science, trends and global diffusion. *International Journal of Nanotechnology and Molecular Computation, 3*(4), 1–23. doi:10.4018/ijnmc.2013100101

El Alouani, H., & Driouchi, A. (2014). The oil and gas sectors, renewable energy, and environmental performance in the Arab world. In A. Driouchi (Ed.), *Knowledge-based economic policy development in the Arab world* (pp. 172–228). Hershey, PA: IGI Global. doi:10.4018/978-1-4666-5210-1.ch008

El-Daoushy, F. (2011). Assessing environment-climate impacts in the Nile basin for decision-making. In *Green technologies: Concepts, methodologies, tools and applications* (pp. 694–712). Hershey, PA: IGI Global. doi:10.4018/978-1-60960-472-1.ch407

Elkarmi, F., & Abu Shikhah, N. (2012). Renewable energy technologies. In *Power system planning technologies and applications: Concepts, solutions and management* (pp. 121–142). Hershey, PA: IGI Global. doi:10.4018/978-1-4666-0173-4.ch008

Ene, C. (2013). Post-consumer waste: Challenges, trends and solutions. *International Journal of Sustainable Economies Management*, 2(3), 19–31. doi:10.4018/ijsem.2013070102

Erdoğdu, M. M., & Karaca, C. (2014). A road map for a domestic wind turbine manufacturing industry in Turkey. In B. Christiansen & M. Basilgan (Eds.), *Economic behavior, game theory, and technology in emerging markets* (pp. 57–90). Hershey, PA: IGI Global. doi:10.4018/978-1-4666-4745-9.ch005

Espiritu, J. F., & Ituarte-Villarreal, C. M. (2013). Wind farm layout optimization using a viral systems algorithm. *International Journal of Applied Evolutionary Computation*, 4(4), 27–40. doi:10.4018/ijaec.2013100102

Evangelista, P., Huge-Brodin, M., Isaksson, K., & Sweeney, E. (2013). Purchasing green transport and logistics services: Implications from the environmental sustainability attitude of 3PLs. In D. Folinas (Ed.), *Outsourcing management for supply chain operations and logistics service* (pp. 449–465). Hershey, PA: IGI Global. doi:10.4018/978-1-4666-2008-7.ch026

Evangelista, P., Huge-Brodin, M., Isaksson, K., & Sweeney, E. (2014). Purchasing green transport and logistics services: Implications from the environmental sustainability attitude of 3PLs. In *Sustainable practices: Concepts, methodologies, tools and applications* (pp. 86–102). Hershey, PA: IGI Global. doi:10.4018/978-1-4666-4852-4.ch005

Ewald, J. A., Sharp, R. J., Beja, P., & Kenward, R. (2013). Pan-European survey and database of environmental assessment factors. In J. Papathanasiou, B. Manos, S. Arampatzis, & R. Kenward (Eds.), *Transactional environmental support system design: Global solutions* (pp. 97–119). Hershey, PA: IGI Global. doi:10.4018/978-1-4666-2824-3.ch006

Fann, J., & Rakas, J. (2011). Greener transportation infrastructure: Theoretical concepts for the environmental evaluation of airports. In Z. Luo (Ed.), *Green finance and sustainability: Environmentally-aware business models and technologies* (pp. 394–421). Hershey, PA: IGI Global. doi:10.4018/978-1-60960-531-5.ch021

Fann, J., & Rakas, J. (2013). Methodology for environmental sustainability evaluation of airport development alternatives. *International Journal of Applied Logistics, 4*(4), 8–31. doi:10.4018/ijal.2013100102

Fara, L., & Yamaguchi, M. (2013). Prospects and strategy of development for advanced solar cells. In L. Fara & M. Yamaguchi (Eds.), *Advanced solar cell materials, technology, modeling, and simulation* (pp. 287–296). Hershey, PA: IGI Global. doi:10.4018/978-1-4666-1927-2.ch014

Farmani, R., Savic, D., Henriksen, H., Molina, J., Giordano, R., & Bromley, J. (2011). Evolutionary Bayesian belief networks for participatory water resources management under uncertainty. In *Green technologies: Concepts, methodologies, tools and applications* (pp. 524–539). Hershey, PA: IGI Global. doi:10.4018/978-1-60960-472-1.ch309

Fearnside, P. M. (2013). Climate change as a threat to Brazil's Amazon forest. *International Journal of Social Ecology and Sustainable Development, 4*(3), 1–12. doi:10.4018/jsesd.2013070101

Filipović, V., Roljević, S., & Bekić, B. (2014). Organic production in Serbia: The transition to green economy. In *Sustainable practices: Concepts, methodologies, tools and applications* (pp. 769–785). Hershey, PA: IGI Global. doi:10.4018/978-1-4666-4852-4.ch043

Finardi, U. (2012). Nanosciences and nanotechnologies: Evolution trajectories and disruptive features. In N. Ekekwe & N. Islam (Eds.), *Disruptive technologies, innovation and global redesign: Emerging implications* (pp. 107–126). Hershey, PA: IGI Global. doi:10.4018/978-1-4666-0134-5.ch007

Finardi, U. (2014). Nanosciences and nanotechnologies: Evolution trajectories and disruptive features. In Nanotechnology: Concepts, methodologies, tools, and applications (pp. 1-20). Hershey, PA: IGI Global. doi:10.4018/978-1-4666-5125-8.ch001

Fokaides, P. A. (2012). Towards zero energy buildings (ZEB): The role of environmental technologies. In O. Ercoskun (Ed.), *Green and ecological technologies for urban planning: Creating smart cities* (pp. 93–111). Hershey, PA: IGI Global. doi:10.4018/978-1-61350-453-6.ch006

Fokaides, P. A. (2014). Towards zero energy buildings (ZEB): The role of environmental technologies. In I. Management Association (Ed.), Sustainable practices: Concepts, methodologies, tools and applications (pp. 1742-1761). Hershey, PA: IGI Global. doi:10.4018/978-1-4666-4852-4.ch096

Gadatsch, A. (2011). Corporate environmental management information systems influence of green IT on IT management and IT controlling. In *Green technologies: Concepts, methodologies, tools and applications* (pp. 1408–1420). Hershey, PA: IGI Global. doi:10.4018/978-1-60960-472-1.ch604

Gálvez, J., Parreño, M., Pla, J., Sanchez, J., Gálvez-Llompart, M., Navarro, S., & García-Domenech, R. (2011). Application of molecular topology to the prediction of water quality indices of Alkylphenol pollutants. *International Journal of Chemoinformatics and Chemical Engineering*, *1*(1), 1–11. doi:10.4018/ijcce.2011010101

Gálvez, J., Parreño, M., Pla, J., Sanchez, J., Gálvez-Llompart, M., Navarro, S., & García-Domenech, R. (2013). Application of molecular topology to the prediction of water quality indices of Alkylphenol pollutants. In A. Haghi (Ed.), *Methodologies and applications for chemoinformatics and chemical engineering* (pp. 1–10). Hershey, PA: IGI Global. doi:10.4018/978-1-4666-4010-8.ch001

Garner, N., Lischke, M. D., Siol, A., & Eilks, I. (2014). Learning about sustainability in a non-formal laboratory context for secondary level students: A module on climate change, the ozone hole, and summer smog. In K. Thomas & H. Muga (Eds.), *Handbook of research on pedagogical innovations for sustainable development* (pp. 229–244). Hershey, PA: IGI Global. doi:10.4018/978-1-4666-5856-1.ch012

Gaurina-Medjimurec, N., & Pasic, B. (2014). CO2 underground storage and wellbore integrity. In D. Matanovic, N. Gaurina-Medjimurec, & K. Simon (Eds.), *Risk analysis for prevention of hazardous situations in petroleum and natural gas engineering* (pp. 322–357). Hershey, PA: IGI Global. doi:10.4018/978-1-4666-4777-0.ch015

Ghosh, N., & Goswami, A. (2014). Biofuels and renewables: Implications for people, planet, policies, politics. In *Sustainability science for social, economic, and environmental development* (pp. 64–87). Hershey, PA: IGI Global. doi:10.4018/978-1-4666-4995-8.ch007

Ghosh, N., & Goswami, A. (2014). Biofuel sustainability and transition pathways. In *Sustainability science for social, economic, and environmental development* (pp. 88–95). Hershey, PA: IGI Global. doi:10.4018/978-1-4666-4995-8.ch008

Ghosh, N., & Goswami, A. (2014). Economics, environmental policy, trade and sustainability. In *Sustainability science for social, economic, and environmental development* (pp. 246–268). Hershey, PA: IGI Global. doi:10.4018/978-1-4666-4995-8.ch016

Ghosh, N., & Goswami, A. (2014). Energy and emission linkages from the three wheeler autorickshaws of Kolkata: An exploratory analysis of the impact on economic, environmental, social dimensions of sustainability. In *Sustainability science for social, economic, and environmental development* (pp. 221–245). Hershey, PA: IGI Global. doi:10.4018/978-1-4666-4995-8.ch015

Ghosh, N., & Goswami, A. (2014). Labour observatories for agricultural policymaking and sustainable development. In *Sustainability science for social, economic, and environmental development* (pp. 56–63). Hershey, PA: IGI Global. doi:10.4018/978-1-4666-4995-8.ch006

Ghosh, N., & Goswami, A. (2014). On value and price of environmental resources. In *Sustainability science for social, economic, and environmental development* (pp. 24–32). Hershey, PA: IGI Global. doi:10.4018/978-1-4666-4995-8.ch002

Ghosh, N., & Goswami, A. (2014). Story of live discussion in autos of Delhi: What do they say about sustainability? In *Sustainability science for social, economic, and environmental development* (pp. 216–220). Hershey, PA: IGI Global. doi:10.4018/978-1-4666-4995-8.ch014

Ghosh, N., & Goswami, A. (2014). Two first generation biofuel (biodiesel, bioethanol) and sustainability: Some other realities for India and trade patterns. In *Sustainability science for social, economic, and environmental development* (pp. 174–208). Hershey, PA: IGI Global. doi:10.4018/978-1-4666-4995-8.ch012

Ghosh, N., & Goswami, A. (2014). Valuation and market-based pricing of economic and ecosystem services of water resources. In *Sustainability science for social, economic, and environmental development* (pp. 96–132). Hershey, PA: IGI Global. doi:10.4018/978-1-4666-4995-8.ch009

Ghosh, N., & Goswami, A. (2014). Water scarcity and conflicts: Can water futures exchange in South Asia provide the answer? In *Sustainability science for social, economic, and environmental development* (pp. 147–173). Hershey, PA: IGI Global. doi:10.4018/978-1-4666-4995-8.ch011

Gil, J., Díaz, L., Granell, C., & Huerta, J. (2013). Open source based deployment of environmental data into geospatial information infrastructures. In I. Management Association (Ed.), Geographic information systems: Concepts, methodologies, tools, and applications (pp. 952-969). Hershey, PA: IGI Global. doi:10.4018/978-1-4666-2038-4.ch059

Gill, L., Hathway, E. A., Lange, E., Morgan, E., & Romano, D. (2013). Coupling real-time 3D landscape models with microclimate simulations. *International Journal of E-Planning Research*, 2(1), 1–19. doi:10.4018/ijepr.2013010101

Giuliani, G., Ray, N., Schwarzer, S., De Bono, A., Peduzzi, P., Dao, H., ... Lehmann, A. (2011). Sharing environmental data through GEOSS. *International Journal of Applied Geospatial Research*, 2(1), 1–17. doi:10.4018/jagr.2011010101

Giuliani, G., Ray, N., Schwarzer, S., De Bono, A., Peduzzi, P., Dao, H., ... Lehmann, A. (2013). Sharing environmental data through GEOSS. In *Geographic information systems: Concepts, methodologies, tools, and applications* (pp. 1260–1275). Hershey, PA: IGI Global. doi:10.4018/978-1-4666-2038-4.ch076

Godbole, N. (2011). E-waste management: Challenges and issues. In B. Unhelkar (Ed.), *Handbook of research on green ICT: Technology, business and social perspectives* (pp. 480–505). Hershey, PA: IGI Global. doi:10.4018/978-1-61692-834-6.ch035

Goel, A., Tiwary, A., & Schmidt, H. (2011). Approaches and initiatives to green IT strategy in business. In *Green technologies: Concepts, methodologies, tools and applications* (pp. 1361–1375). Hershey, PA: IGI Global. doi:10.4018/978-1-60960-472-1.ch601

Granit, J. J., King, R. M., & Noël, R. (2013). Strategic environmental assessment as a tool to develop power in transboundary water basin settings. In E. Carayannis (Ed.), *Creating a sustainable ecology using technology-driven solutions* (pp. 269–281). Hershey, PA: IGI Global. doi:10.4018/978-1-4666-3613-2.ch018

Gräuler, M., Teuteberg, F., Mahmoud, T., & Gómez, J. M. (2013). Requirements prioritization and design considerations for the next generation of corporate environmental management information systems: A foundation for innovation. *International Journal of Information Technologies and Systems Approach*, 6(1), 98–116. doi:10.4018/jitsa.2013010106

Gräuler, M., Teuteberg, F., Mahmoud, T., & Gómez, J. M. (2013). Requirements prioritization and design considerations for the next generation of corporate environmental management information systems: A foundation for innovation. *International Journal of Information Technologies and Systems Approach*, 6(1), 98–116. doi:10.4018/jitsa.2013010106

Greenlee, B., & Daim, T. (2011). Building a sustainable regional eco system for green technologies: Case of cellulosic ethanol in Oregon. In Z. Luo (Ed.), *Green finance and sustainability: Environmentally-aware business models and technologies* (pp. 535–568). Hershey, PA: IGI Global. doi:10.4018/978-1-60960-531-5.ch028

Greenlee, B., & Daim, T. (2013). Building a sustainable regional eco system for green technologies: Case of cellulosic ethanol in Oregon. In I. Management Association (Ed.), *Small and medium enterprises: Concepts, methodologies, tools, and applications* (pp. 993-1025). Hershey, PA: IGI Global. doi:10.4018/978-1-4666-3886-0.ch049

Grigoroudis, E., Kouikoglou, V. S., & Phillis, Y. A. (2012). Approaches for measuring sustainability. In P. Olla (Ed.), *Global sustainable development and renewable energy systems* (pp. 101–130). Hershey, PA: IGI Global. doi:10.4018/978-1-4666-1625-7.ch006

Grigoroudis, E., Kouikoglou, V. S., & Phillis, Y. A. (2014). Approaches for measuring sustainability. In *Sustainable practices: Concepts, methodologies, tools and applications* (pp. 158–184). Hershey, PA: IGI Global. doi:10.4018/978-1-4666-4852-4.ch009

Guangming, L., & Zhaofeng, A. (2013). Empirical study on the correlations of environmental pollution, human capital, and economic growth: Based on the 1990-2007 data in Guangdong China. In P. Ordóñez de Pablos (Ed.), *Green technologies and business practices: An IT approach* (pp. 128–137). Hershey, PA: IGI Global. doi:10.4018/978-1-4666-1972-2.ch006

Gupta, A. K., Chakraborty, A., Giri, S., Subramanian, V., & Chattaraj, P. (2011). Toxicity of halogen, sulfur and chlorinated aromatic compounds: A quantitative-structure-toxicity-relationship (QSTR). *International Journal of Chemoinformatics and Chemical Engineering*, *1*(1), 61–74. doi:10.4018/ijcce.2011010105

Gupta, A. K., Chakraborty, A., Giri, S., Subramanian, V., & Chattaraj, P. (2013). Toxicity of halogen, sulfur and chlorinated aromatic compounds: A quantitative-structure-toxicity-relationship (QSTR). In A. Haghi (Ed.), *Methodologies and applications for chemoinformatics and chemical engineering* (pp. 60–73). Hershey, PA: IGI Global. doi:10.4018/978-1-4666-4010-8.ch005

Habala, O., Šeleng, M., Tran, V., Šimo, B., & Hluchý, L. (2012). Mining environmental data in the ADMIRE project using new advanced methods and tools. In N. Bessis (Ed.), *Technology integration advancements in distributed systems and computing* (pp. 296–308). Hershey, PA: IGI Global. doi:10.4018/978-1-4666-0906-8.ch018

Hájek, P., & Olej, V. (2011). Air quality assessment by neural networks. In V. Olej, I. Obršálová, & J. Krupka (Eds.), *Environmental modeling for sustainable regional development: System approaches and advanced methods* (pp. 91–117). Hershey, PA: IGI Global. doi:10.4018/978-1-60960-156-0.ch005

Hall, C., Easley, R., Howard, J., & Halfhide, T. (2013). The role of authentic science research and education outreach in increasing community resilience: Case studies using informal education to address ocean acidification and healthy soils. In H. Muga & K. Thomas (Eds.), *Cases on the diffusion and adoption of sustainable development practices* (pp. 376–402). Hershey, PA: IGI Global. doi:10.4018/978-1-4666-2842-7.ch014

Hall, G. M., & Howe, J. (2013). The drivers for a sustainable chemical manufacturing industry. In *Industrial engineering: Concepts, methodologies, tools, and applications* (pp. 1659–1679). Hershey, PA: IGI Global. doi:10.4018/978-1-4666-1945-6.ch088

Hashemi, M., & O'Connell, E. (2013). Science and water policy interface: An integrated methodological framework for developing decision support systems (DSSs). In *Data mining: Concepts, methodologies, tools, and applications* (pp. 405–434). Hershey, PA: IGI Global. doi:10.4018/978-1-4666-2455-9.ch020

Heck, M., & Schmidt, G. (2013). Lot-size planning with non-linear cost functions supporting environmental sustainability. In K. Ganesh & S. Anbuudayasankar (Eds.), *International and interdisciplinary studies in green computing* (pp. 226–231). Hershey, PA: IGI Global. doi:10.4018/978-1-4666-2646-1.ch016

Herold, S., & Sawada, M. C. (2013). A review of geospatial information technology for natural disaster management in developing countries. In *Geographic information systems: Concepts, methodologies, tools, and applications* (pp. 175–215). Hershey, PA: IGI Global. doi:10.4018/978-1-4666-2038-4.ch014

Higginson, N., & Vredenburg, H. (2012). Finding the sweet spot of sustainability in the energy sector: A systems approach to managing the canadian oil sands. In M. Tortora (Ed.), *Sustainable systems and energy management at the regional level: Comparative approaches* (pp. 184–201). Hershey, PA: IGI Global. doi:10.4018/978-1-61350-344-7.ch010

Hilty, L. M. (2011). Information and communication technologies for a more sustainable world. In D. Haftor & A. Mirijamdotter (Eds.), *Information and communication technologies, society and human beings: Theory and framework (festschrift in honor of Gunilla Bradley)* (pp. 410–418). Hershey, PA: IGI Global. doi:10.4018/978-1-60960-057-0.ch033

Hin, L. T., & Subramaniam, R. (2012). Use of policy instruments to promote sustainable energy practices and implications for the environment: Experiences from Singapore. In M. Tortora (Ed.), *Sustainable systems and energy management at the regional level: Comparative approaches* (pp. 219–235). Hershey, PA: IGI Global. doi:10.4018/978-1-61350-344-7.ch012

Hrncevic, L. (2014). Petroleum industry environmental performance and risk. In D. Matanovic, N. Gaurina-Medjimurec, & K. Simon (Eds.), *Risk analysis for prevention of hazardous situations in petroleum and natural gas engineering* (pp. 358–387). Hershey, PA: IGI Global. doi:10.4018/978-1-4666-4777-0.ch016

Hsiao, S., Chen, D., Yang, C., Huang, H., Lu, Y., Huang, H., ... Lin, Y. (2013). Chemical-free and reusable cellular analysis: Electrochemical impedance spectroscopy with a transparent ITO culture chip. *International Journal of Technology and Human Interaction, 8*(3), 1–9. doi:10.4018/jthi.2012070101

Hunter, J., Becker, P., Alabri, A., van Ingen, C., & Abal, E. (2011). Using ontologies to relate resource management actions to environmental monitoring data in South East Queensland. *International Journal of Agricultural and Environmental Information Systems, 2*(1), 1–19. doi:10.4018/jaeis.2011010101

Imbrenda, V., D'Emilio, M., Lanfredi, M., Ragosta, M., & Simoniello, T. (2013). Indicators of land degradation vulnerability due to anthropic factors: Tools for an efficient planning. In G. Borruso, S. Bertazzon, A. Favretto, B. Murgante, & C. Torre (Eds.), *Geographic information analysis for sustainable development and economic planning: New technologies* (pp. 87–101). Hershey, PA: IGI Global. doi:10.4018/978-1-4666-1924-1.ch006

Imbrenda, V., D'Emilio, M., Lanfredi, M., Ragosta, M., & Simoniello, T. (2014). Indicators of land degradation vulnerability due to anthropic factors: Tools for an efficient planning. In I. Management Association (Ed.), Sustainable practices: Concepts, methodologies, tools and applications (pp. 1400-1413). Hershey, PA: IGI Global. doi:10.4018/978-1-4666-4852-4.ch078

Iojă, C., Niță, M. R., & Stupariu, I. G. (2014). Resource conservation: Key elements in sustainable rural development. In Z. Andreopoulou, V. Samathrakis, S. Louca, & M. Vlachopoulou (Eds.), *E-innovation for sustainable development of rural resources during global economic crisis* (pp. 80–97). Hershey, PA: IGI Global. doi:10.4018/978-1-4666-4550-9.ch008

Ioja, C., Rozylowicz, L., Patroescu, M., Nita, M., & Onose, D. (2011). Agriculture and conservation in the Natura 2000 network: A sustainable development approach of the European Union. In Z. Andreopoulou, B. Manos, N. Polman, & D. Viaggi (Eds.), *Agricultural and environmental informatics, governance and management: Emerging research applications* (pp. 339–358). Hershey, PA: IGI Global. doi:10.4018/978-1-60960-621-3.ch018

Ioja, C., Rozylowicz, L., Patroescu, M., Nita, M., & Onose, D. (2013). Agriculture and conservation in the Natura 2000 network: A sustainable development approach of the European Union. In *Geographic information systems: Concepts, methodologies, tools, and applications* (pp. 1276–1296). Hershey, PA: IGI Global. doi:10.4018/978-1-4666-2038-4.ch077

Ip-Soo-Ching, J. M., & Zyngier, S. (2014). The rise of "environmental sustainability knowledge" in business strategy and entrepreneurship: An IT-enabled knowledge-based view of tourism operators. In P. Ordóñez de Pablos (Ed.), International business strategy and entrepreneurship: An information technology perspective (pp. 23-40). Hershey, PA: IGI Global. doi:10.4018/978-1-4666-4753-4.ch002

Ivask, M., Aruvee, E., & Piirimäe, K. (2013). Database of environmental decision support tools. In J. Papathanasiou, B. Manos, S. Arampatzis, & R. Kenward (Eds.), *Transactional environmental support system design: Global solutions* (pp. 70–96). Hershey, PA: IGI Global. doi:10.4018/978-1-4666-2824-3.ch005

Jacobsson, M., Linde, A., & Linderoth, H. (2011). The relation between ICT and environmental management practice in a construction company. In *Green technologies: Concepts, methodologies, tools and applications* (pp. 1099–1117). Hershey, PA: IGI Global. doi:10.4018/978-1-60960-472-1.ch430

Jafari, M. (2013). Challenges in climate change and environmental crisis: Impacts of aviation industry on human, urban and natural environments. *International Journal of Space Technology Management and Innovation*, 3(2), 24–46. doi:10.4018/ijstmi.2013070102

Jain, H. (2011). Green ICT organizational implementations and workplace relationships. In B. Unhelkar (Ed.), *Handbook of research on green ICT: Technology, business and social perspectives* (pp. 146–168). Hershey, PA: IGI Global. doi:10.4018/978-1-61692-834-6.ch010

Jamous, N. (2013). Light-weight composite environmental performance indicators (LWC-EPI): A new approach for environmental management information systems (EMIS). *International Journal of Information Technologies and Systems Approach*, 6(1), 20–38. doi:10.4018/jitsa.2013010102

Jan, Y., Lin, M., Shiao, K., Wei, C., Huang, L., & Sung, Q. (2013). Development of an evaluation instrument for green building literacy among college students in Taiwan. *International Journal of Technology and Human Interaction*, 8(3), 31–45. doi:10.4018/jthi.2012070104

Jarmoszko, A., D'Onofrio, M., Lee-Partridge, J. E., & Petkova, O. (2013). Evaluating sustainability and greening methods: A conceptual model for information technology management. *International Journal of Applied Logistics*, 4(3), 1–13. doi:10.4018/jal.2013070101

Jena, R. K. (2013). Green computing to green business. In P. Ordóñez de Pablos (Ed.), *Green technologies and business practices: An IT approach* (pp. 138–150). Hershey, PA: IGI Global. doi:10.4018/978-1-4666-1972-2.ch007

Jinturkar, A. M., & Deshmukh, S. S. (2013). Sustainable development by rural energy resources allocation in India: A fuzzy goal programming approach. *International Journal of Energy Optimization and Engineering*, *2*(1), 37–49. doi:10.4018/ijeoe.2013010103

Jirava, P., & Obršálová, I. (2011). Modeling the effects of the quality of the environment on the health of a selected population. In V. Olej, I. Obršálová, & J. Krupka (Eds.), *Environmental modeling for sustainable regional development: System approaches and advanced methods* (pp. 344–365). Hershey, PA: IGI Global. doi:10.4018/978-1-60960-156-0.ch017

Jonoski, A., & Evers, M. (2013). Sociotechnical framework for participatory flood risk management via collaborative modeling. *International Journal of Information Systems and Social Change*, *4*(2), 1–16. doi:10.4018/jissc.2013040101

Joshi, P. K., & Priyanka, N. (2011). Geo-informatics for land use and biodiversity studies. In Y. Trisurat, R. Shrestha, & R. Alkemade (Eds.), *Land use, climate change and biodiversity modeling: Perspectives and applications* (pp. 52–77). Hershey, PA: IGI Global. doi:10.4018/978-1-60960-619-0.ch003

Joshi, P. K., & Priyanka, N. (2013). Geo-informatics for land use and biodiversity studies. In I. Management Association (Ed.), Geographic information systems: Concepts, methodologies, tools, and applications (pp. 1913-1939). Hershey, PA: IGI Global. doi:10.4018/978-1-4666-2038-4.ch114

Júnior, R., Rigitano, R., & Boesten, J. (2011). Pesticide leaching models in a Brazilian agricultural field scenario. In H. do Prado, A. Barreto Luiz, & H. Filho (Eds.), *Computational methods for agricultural research: Advances and applications* (pp. 266–295). Hershey, PA: IGI Global. doi:10.4018/978-1-61692-871-1.ch013

Kader, A. S., & Olanrewaju, O. S. (2014). River transportation master plan study for environmental enhancement. In O. Olanrewaju, A. Saharuddin, A. Ab Kader, & W. Wan Nik (Eds.), *Marine technology and sustainable development: Green innovations* (pp. 178–184). Hershey, PA: IGI Global. doi:10.4018/978-1-4666-4317-8.ch011

Kamaja, C. K., Rajaperumal, M., Boukherroub, R., & Shelke, M. V. (2014). Silicon nanostructures-graphene nanocomposites: Efficient materials for energy conversion and storage. In M. Bououdina & J. Davim (Eds.), *Handbook of research on nanoscience, nanotechnology, and advanced materials* (pp. 176–195). Hershey, PA: IGI Global. doi:10.4018/978-1-4666-5824-0.ch009

Kaplan, A. (2012). "Green infrastructure" concept as an effective medium to manipulating sustainable urban development. In O. Ercoskun (Ed.), *Green and ecological technologies for urban planning: Creating smart cities* (pp. 234–254). Hershey, PA: IGI Global. doi:10.4018/978-1-61350-453-6.ch013

Kašparová, M., & Krupka, J. (2011). Air quality modeling and metamodeling approach. In V. Olej, I. Obršálová, & J. Krupka (Eds.), *Environmental modeling for sustainable regional development: System approaches and advanced methods* (pp. 144–161). Hershey, PA: IGI Global. doi:10.4018/978-1-60960-156-0.ch007

Kenward, R., Casey, N. M., Walls, S. S., Dick, J. M., Smith, R., & Turner, S. L. ... Sharp, R. J. (2013). Pan-European analysis of environmental assessment processes. In J. Papathanasiou, B. Manos, S. Arampatzis, & R. Kenward (Eds.), Transactional environmental support system design: Global solutions (pp. 120-133). Hershey, PA: IGI Global. doi:10.4018/978-1-4666-2824-3.ch007

Kokkinakis, A., & Andreopoulou, Z. (2011). E-governance and management of inland water ecosystems using time-series analysis of fishery production. In Z. Andreopoulou, B. Manos, N. Polman, & D. Viaggi (Eds.), *Agricultural and environmental informatics, governance and management: Emerging research applications* (pp. 318–338). Hershey, PA: IGI Global. doi:10.4018/978-1-60960-621-3.ch017

Kongar, E. A., & Rosentrater, K. (2013). Data envelopment analysis approach to compare the environmental efficiency of energy utilization. In K. Ganesh & S. Anbuudayasankar (Eds.), *International and interdisciplinary studies in green computing* (pp. 273–288). Hershey, PA: IGI Global. doi:10.4018/978-1-4666-2646-1.ch020

Kosaka, M., Yabutani, T., & Zhang, Q. (2014). A value co-creation model for energy-saving service business using inverters. In M. Kosaka & K. Shirahada (Eds.), *Progressive trends in knowledge and system-based science for service innovation* (pp. 292–306). Hershey, PA: IGI Global. doi:10.4018/978-1-4666-4663-6.ch016

Kram, T., & Stehfest, E. (2011). Integrated modeling of global environmental change (IMAGE). In Y. Trisurat, R. Shrestha, & R. Alkemade (Eds.), *Land use, climate change and biodiversity modeling: Perspectives and applications* (pp. 104–118). Hershey, PA: IGI Global. doi:10.4018/978-1-60960-619-0.ch005

La Greca, P., La Rosa, D., Martinico, F., & Privitera, R. (2013). Land cover analysis for evapotranspiration assessment in Catania metropolitan region. In G. Borruso, S. Bertazzon, A. Favretto, B. Murgante, & C. Torre (Eds.), *Geographic information analysis for sustainable development and economic planning: New technologies* (pp. 102–114). Hershey, PA: IGI Global. doi:10.4018/978-1-4666-1924-1.ch007

Laike, Y., & Chun, L. (2012). China-European Union trade and global warming. In E. Carayannis (Ed.), *Sustainable policy applications for social ecology and development* (pp. 18–28). Hershey, PA: IGI Global. doi:10.4018/978-1-4666-1586-1.ch003

Laing, R., Bennadji, A., & Gray, D. (2013). Traffic control and CO2 reduction: Utilisation of virtual modelling within university estates master planning. *International Journal of E-Planning Research*, *2*(1), 43–57. doi:10.4018/ijepr.2013010103

Lam, J. C., & Hills, P. (2011). Promoting technological environmental innovations: What is the role of environmental regulation? In Z. Luo (Ed.), *Green finance and sustainability: Environmentally-aware business models and technologies* (pp. 56–73). Hershey, PA: IGI Global. doi:10.4018/978-1-60960-531-5.ch003

Lam, J. C., & Hills, P. (2012). Transition to low-carbon hydrogen economy in America: The role of transition management. In Z. Luo (Ed.), *Advanced analytics for green and sustainable economic development: Supply chain models and financial technologies* (pp. 92–111). Hershey, PA: IGI Global. doi:10.4018/978-1-61350-156-6.ch007

Lam, J. C., & Hills, P. (2013). Promoting technological environmental innovations: The role of environmental regulation. In Z. Luo (Ed.), *Technological solutions for modern logistics and supply chain management* (pp. 230–247). Hershey, PA: IGI Global. doi:10.4018/978-1-4666-2773-4.ch015

Lee, S., Yigitcanlar, T., Egodawatta, P., & Goonetilleke, A. (2011). Sustainable water provision. In *Green technologies: Concepts, methodologies, tools and applications* (pp. 1768–1781). Hershey, PA: IGI Global. doi:10.4018/978-1-60960-472-1.ch714

Lee, Y. M., An, L., Liu, F., Horesh, R., Chae, Y. T., & Zhang, R. (2014). Analytics for smarter buildings. *International Journal of Business Analytics*, *1*(1), 1–15. doi:10.4018/ijban.2014010101

Leff, E. (2012). Environmental rationality: Innovation in thinking for sustainability. In F. Nobre, D. Walker, & R. Harris (Eds.), *Technological, managerial and organizational core competencies: Dynamic innovation and sustainable development* (pp. 1–17). Hershey, PA: IGI Global. doi:10.4018/978-1-61350-165-8.ch001

Leff, E. (2014). Environmental rationality: Innovation in thinking for sustainability. In *Sustainable practices: Concepts, methodologies, tools and applications* (pp. 1–17). Hershey, PA: IGI Global. doi:10.4018/978-1-4666-4852-4.ch001

Lefley, F., & Sarkis, J. (2011). A pragmatic profile approach to evaluating environmental sustainability investment decisions. In Z. Luo (Ed.), *Green finance and sustainability: Environmentally-aware business models and technologies* (pp. 321–332). Hershey, PA: IGI Global. doi:10.4018/978-1-60960-531-5.ch017

Li, H., & Zhang, X. (2012). Study on environmental tax: A case of China. In D. Ura & P. Ordóñez de Pablos (Eds.), *Advancing technologies for Asian business and economics: Information management developments* (pp. 207–219). Hershey, PA: IGI Global. doi:10.4018/978-1-4666-0276-2.ch016

Li, X., Ortiz, P., Kuczenski, B., Franklin, D., & Chong, F. T. (2012). Mitigating the environmental impact of smartphones with device reuse. In W. Hu & N. Kaabouch (Eds.), *Sustainable ICTs and management systems for green computing* (pp. 252–282). Hershey, PA: IGI Global. doi:10.4018/978-1-4666-1839-8.ch011

Li, X., Ortiz, P. J., Browne, J., Franklin, D., Oliver, J. Y., Geyer, R., ... Chong, F. T. (2012). A study of reusing smartphones to augment elementary school education. *International Journal of Handheld Computing Research, 3*(2), 73–92. doi:10.4018/jhcr.2012040105

Lin, C., Chu, L., & Hsu, H. (2013). Study on the performance and exhaust emissions of motorcycle engine fuelled with hydrogen-gasoline compound fuel. *International Journal of Technology and Human Interaction, 8*(3), 69–81. doi:10.4018/jthi.2012070107

Lingarchani, A. (2011). Environmental challenges in mobile services. In B. Unhelkar (Ed.), *Handbook of research on green ICT: Technology, business and social perspectives* (pp. 355–363). Hershey, PA: IGI Global. doi:10.4018/978-1-61692-834-6.ch025

Lingarchani, A. (2012). Environmental challenges in mobile services. In *Wireless technologies: Concepts, methodologies, tools and applications* (pp. 1891–1899). Hershey, PA: IGI Global. doi:10.4018/978-1-61350-101-6.ch710

Loeser, F., Erek, K., & Zarnekow, R. (2013). Green IT strategies: A conceptual framework for the alignment of information technology and corporate sustainability strategy. In P. Ordóñez de Pablos (Ed.), *Green technologies and business practices: An IT approach* (pp. 58–95). Hershey, PA: IGI Global. doi:10.4018/978-1-4666-1972-2.ch004

Loi, N. K. (2013). Sustainable land use and watershed management in response to climate change impacts: Overview and proposed research techniques. In *Geographic information systems: Concepts, methodologies, tools, and applications* (pp. 2080–2101). Hershey, PA: IGI Global. doi:10.4018/978-1-4666-2038-4.ch124

Lucignano, C., Squeo, E. A., Guglielmotti, A., & Quadrini, F. (2013). Recycling of waste epoxy-polyester powders for foam production. In J. Davim (Ed.), *Dynamic methods and process advancements in mechanical, manufacturing, and materials engineering* (pp. 91–101). Hershey, PA: IGI Global. doi:10.4018/978-1-4666-1867-1.ch005

Mahbub, P., Ayoko, G., Egodawatta, P., Yigitcanlar, T., & Goonetilleke, A. (2011). Traffic and climate change impacts on water quality. In I. Management Association (Ed.), Green technologies: Concepts, methodologies, tools and applications (pp. 1804-1823). Hershey, PA: IGI Global. doi:10.4018/978-1-60960-472-1.ch716

Maillé, E., & Espinasse, B. (2011). Pyroxene: A territorial decision support system based on spatial simulators integration for forest fire risk management. *International Journal of Agricultural and Environmental Information Systems*, 2(2), 52–72. doi:10.4018/jaeis.2011070104

Maillé, E., & Espinasse, B. (2012). Pyroxene: A territorial decision support system based on spatial simulators integration for forest fire risk management. In P. Papajorgji & F. Pinet (Eds.), *New technologies for constructing complex agricultural and environmental systems* (pp. 244–264). Hershey, PA: IGI Global. doi:10.4018/978-1-4666-0333-2.ch014

Mallios, Z. (2012). Irrigation water valuation using spatial hedonic models in GIS environment. In J. Wang (Ed.), *Societal impacts on information systems development and applications* (pp. 308–320). Hershey, PA: IGI Global. doi:10.4018/978-1-4666-0927-3.ch020

Manikas, I., Ieromonachou, P., & Bochtis, D. (2014). Environmental sustainability initiatives in the agrifood supply chain. In Z. Andreopoulou, V. Samathrakis, S. Louca, & M. Vlachopoulou (Eds.), *E-innovation for sustainable development of rural resources during global economic crisis* (pp. 221–232). Hershey, PA: IGI Global. doi:10.4018/978-1-4666-4550-9.ch016

Manou, D., & Papathanasiou, J. (2013). Exploring the development of new tourism activities in the municipality of Kerkini by using the area's natural resources sustainably, municipality of Kerkini, Greece. In J. Papathanasiou, B. Manos, S. Arampatzis, & R. Kenward (Eds.), *Transactional environmental support system design: Global solutions* (pp. 172–175). Hershey, PA: IGI Global. doi:10.4018/978-1-4666-2824-3.ch012

Maragkogianni, A., Papaefthimiou, S., & Zopounidis, C. (2013). Emissions trading schemes in the transportation sector. In A. Jean-Vasile, T. Adrian, J. Subic, & D. Dusmanescu (Eds.), *Sustainable technologies, policies, and constraints in the green economy* (pp. 269–289). Hershey, PA: IGI Global. doi:10.4018/978-1-4666-4098-6.ch015

Maragkogianni, A., Papaefthimiou, S., & Zopounidis, C. (2014). Emissions trading schemes in the transportation sector. In *Sustainable practices: Concepts, methodologies, tools and applications* (pp. 65–85). Hershey, PA: IGI Global. doi:10.4018/978-1-4666-4852-4.ch004

Marino, D. J., Castro, E. A., Massolo, L., Mueller, A., Herbarth, O., & Ronco, A. E. (2011). Characterization of polycyclic aromatic hydrocarbon profiles by multivariate statistical analysis. *International Journal of Chemoinformatics and Chemical Engineering*, *1*(2), 1–14. doi:10.4018/ijcce.2011070101

Marino, D. J., Castro, E. A., Massolo, L., Mueller, A., Herbarth, O., & Ronco, A. E. (2013). Characterization of polycyclic aromatic hydrocarbon profiles by multivariate statistical analysis. In A. Haghi (Ed.), *Methodologies and applications for chemoinformatics and chemical engineering* (pp. 102–116). Hershey, PA: IGI Global. doi:10.4018/978-1-4666-4010-8.ch008

Marshall, A. (2011). The middle ground for nuclear waste management: Social and ethical aspects of shallow storage. *International Journal of Technoethics*, *2*(2), 1–13. doi:10.4018/jte.2011040101

Mbzibain, A. (2013). The effect of farmer capacities, farm business resources and perceived support of family, friends and associational networks on intentions to invest in renewable energy ventures in the UK. *International Journal of Applied Behavioral Economics*, *2*(3), 43–58. doi:10.4018/ijabe.2013070104

Mbzibain, A. (2014). The effect of farmer capacities, farm business resources and perceived support of family, friends and associational networks on intentions to invest in renewable energy ventures in the UK. In *Sustainable practices: Concepts, methodologies, tools and applications* (pp. 1072–1088). Hershey, PA: IGI Global. doi:10.4018/978-1-4666-4852-4.ch059

McKnight, K. P., Messina, J. P., Shortridge, A. M., Burns, M. D., & Pigozzi, B. W. (2011). Using volunteered geographic information to assess the spatial distribution of West Nile Virus in Detroit, Michigan. *International Journal of Applied Geospatial Research*, *2*(3), 72–85. doi:10.4018/jagr.2011070105

Mengel, M. A. (2011). Constructing an experience in a virtual green home. In G. Vincenti & J. Braman (Eds.), *Multi-user virtual environments for the classroom: Practical approaches to teaching in virtual worlds* (pp. 285–301). Hershey, PA: IGI Global. doi:10.4018/978-1-60960-545-2.ch018

Miidla, P. (2011). Data envelopment analysis in environmental technologies. In V. Olej, I. Obršálová, & J. Krupka (Eds.), *Environmental modeling for sustainable regional development: System approaches and advanced methods* (pp. 242–259). Hershey, PA: IGI Global. doi:10.4018/978-1-60960-156-0.ch012

Miidla, P. (2013). Data envelopment analysis in environmental technologies. In *Industrial engineering: Concepts, methodologies, tools, and applications* (pp. 625–642). Hershey, PA: IGI Global. doi:10.4018/978-1-4666-1945-6.ch036

Militano, L., Molinaro, A., Iera, A., & Petkovics, Á. (2013). A game theoretic approach to guarantee fairness in cooperation among green mobile network operators. *International Journal of Business Data Communications and Networking*, *9*(3), 1–15. doi:10.4018/jbdcn.2013070101

Miller, W., & Birkeland, J. (2011). Green energy. In *Green technologies: Concepts, methodologies, tools and applications* (pp. 1–16). Hershey, PA: IGI Global. doi:10.4018/978-1-60960-472-1.ch101

Miralles, A., Pinet, F., & Bédard, Y. (2012). Describing spatio-temporal phenomena for environmental system development: An overview of today's needs and solutions. In P. Papajorgji & F. Pinet (Eds.), *New technologies for constructing complex agricultural and environmental systems* (pp. 211–226). Hershey, PA: IGI Global. doi:10.4018/978-1-4666-0333-2.ch012

Misso, R. (2011). Sustainable governance in the integrated system "environment-agriculture–health" through ICTs. In Z. Andreopoulou, B. Manos, N. Polman, & D. Viaggi (Eds.), *Agricultural and environmental informatics, governance and management: Emerging research applications* (pp. 87–101). Hershey, PA: IGI Global. doi:10.4018/978-1-60960-621-3.ch005

Mitroi, M. R., Fara, L., & Moraru, A. G. (2013). Organic solar cells modeling and simulation. In L. Fara & M. Yamaguchi (Eds.), *Advanced solar cell materials, technology, modeling, and simulation* (pp. 120–137). Hershey, PA: IGI Global. doi:10.4018/978-1-4666-1927-2.ch008

Mochal, T., & Krasnoff, A. (2013). GreenPM®: The basic principles for applying an environmental dimension to project management. In G. Silvius & J. Tharp (Eds.), *Sustainability integration for effective project management* (pp. 39–57). Hershey, PA: IGI Global. doi:10.4018/978-1-4666-4177-8.ch003

Montgomery, M. C., & Chakraborty, J. (2013). Social vulnerability to coastal and inland flood hazards: A comparison of GIS-based spatial interpolation methods. *International Journal of Applied Geospatial Research*, *4*(3), 58–79. doi:10.4018/jagr.2013070104

Moreno, I. S., & Xu, J. (2013). Energy-efficiency in cloud computing environments: Towards energy savings without performance degradation. In S. Aljawarneh (Ed.), *Cloud computing advancements in design, implementation, and technologies* (pp. 18–36). Hershey, PA: IGI Global. doi:10.4018/978-1-4666-1879-4.ch002

Morris, J. Z., & Thomas, K. D. (2013). Implementing biosand filters in rural Honduras: A case study of his hands mission international in Copán, Honduras. In H. Muga & K. Thomas (Eds.), *Cases on the diffusion and adoption of sustainable development practices* (pp. 468–496). Hershey, PA: IGI Global. doi:10.4018/978-1-4666-2842-7.ch017

Mu, Z., Jing, L., Xiaohong, Z., Lei, T., Xiao-na, F., & Shan, C. (2011). Study on low-carbon economy model and method of Chinese tourism industry. *International Journal of Applied Logistics*, *2*(2), 69–102. doi:10.4018/jal.2011040105

Mu, Z., Jing, L., Xiaohong, Z., Lei, T., Xiao-na, F., & Shan, C. (2013). Study on low-carbon economy model and method of chinese tourism industry. In Z. Luo (Ed.), *Technological solutions for modern logistics and supply chain management* (pp. 284–317). Hershey, PA: IGI Global. doi:10.4018/978-1-4666-2773-4.ch018

Mudhoo, A., & Lin, Z. (2012). Phytoremediation of nickel: Mechanisms, application and management. In N. Ekekwe & N. Islam (Eds.), *Disruptive technologies, innovation and global redesign: Emerging implications* (pp. 173–195). Hershey, PA: IGI Global. doi:10.4018/978-1-4666-0134-5.ch010

Murugesan, S. (2011). Strategies for greening enterprise IT: Creating business value and contributing to environmental sustainability. In B. Unhelkar (Ed.), *Handbook of research on green ICT: Technology, business and social perspectives* (pp. 51–64). Hershey, PA: IGI Global. doi:10.4018/978-1-61692-834-6.ch004

Nagni, M., & Ventouras, S. (2013). Implementation of UML schema in relational databases: A case of geographic information. *International Journal of Distributed Systems and Technologies*, *4*(4), 50–60. doi:10.4018/ijdst.2013100105

Nair, S. R., & Ndubisi, N. O. (2013). Entrepreneurial values, environmental marketing and customer satisfaction: Conceptualization and propositions. In N. Ndubisi & S. Nwankwo (Eds.), *Enterprise development in SMEs and entrepreneurial firms: Dynamic processes* (pp. 257–269). Hershey, PA: IGI Global. doi:10.4018/978-1-4666-2952-3.ch013

Nourani, V., Roumianfar, S., & Sharghi, E. (2013). Using hybrid ARIMAX-ANN model for simulating rainfall - runoff - sediment process case study: Aharchai Basin, Iran. *International Journal of Applied Metaheuristic Computing, 4*(2), 44–60. doi:10.4018/jamc.2013040104

Obara, S. (2011). Fuel reduction effect of the solar cell and diesel engine hybrid system with a prediction algorithm of solar power generation. In *Green technologies: Concepts, methodologies, tools and applications* (pp. 815–839). Hershey, PA: IGI Global. doi:10.4018/978-1-60960-472-1.ch414

Oktay, D. (2014). Sustainable urbanism revisited: A holistic framework based on tradition and contemporary orientations. In *Sustainable practices: Concepts, methodologies, tools and applications* (pp. 1723–1741). Hershey, PA: IGI Global. doi:10.4018/978-1-4666-4852-4.ch095

Olanrewaju, O. S. (2014). Evolving sustainable green ship technology. In O. Olanrewaju, A. Saharuddin, A. Ab Kader, & W. Wan Nik (Eds.), *Marine technology and sustainable development: Green innovations* (pp. 127–145). Hershey, PA: IGI Global. doi:10.4018/978-1-4666-4317-8.ch008

Olanrewaju, O. S. (2014). Risk requirement for multi-hybrid renewable energy for marine system. In O. Olanrewaju, A. Saharuddin, A. Ab Kader, & W. Wan Nik (Eds.), *Marine technology and sustainable development: Green innovations* (pp. 83–95). Hershey, PA: IGI Global. doi:10.4018/978-1-4666-4317-8.ch005

Olanrewaju, O. S., & Kader, A. S. (2014). Applying the safety and environmental risk and reliability model (SERM) for Malaysian Langat River collision aversion. In O. Olanrewaju, A. Saharuddin, A. Ab Kader, & W. Wan Nik (Eds.), *Marine technology and sustainable development: Green innovations* (pp. 193–225). Hershey, PA: IGI Global. doi:10.4018/978-1-4666-4317-8.ch013

Olej, V., & Hájek, P. (2011). Air quality modeling by fuzzy sets and IF-sets. In V. Olej, I. Obršálová, & J. Krupka (Eds.), *Environmental modeling for sustainable regional development: System approaches and advanced methods* (pp. 118–143). Hershey, PA: IGI Global. doi:10.4018/978-1-60960-156-0.ch006

Omer, A. M. (2012). Renewable energy and sustainable development. In P. Vasant, N. Barsoum, & J. Webb (Eds.), *Innovation in power, control, and optimization: Emerging energy technologies* (pp. 95–136). Hershey, PA: IGI Global. doi:10.4018/978-1-61350-138-2.ch003

Omer, A. M. (2014). Cooling and heating with ground source energy. In *Sustainable practices: Concepts, methodologies, tools and applications* (pp. 261–278). Hershey, PA: IGI Global. doi:10.4018/978-1-4666-4852-4.ch014

Ondieki, C. M. (2013). Hydrology and integrated water resource management for sustainable watershed management in Kenya. In H. Muga & K. Thomas (Eds.), *Cases on the diffusion and adoption of sustainable development practices* (pp. 352–375). Hershey, PA: IGI Global. doi:10.4018/978-1-4666-2842-7.ch013

Ozbakir, B. A. (2012). Urban environmental applications of GIScience: Challenges and new trends. In O. Ercoskun (Ed.), *Green and ecological technologies for urban planning: Creating smart cities* (pp. 192–211). Hershey, PA: IGI Global. doi:10.4018/978-1-61350-453-6.ch011

Ozbakir, B. A. (2014). Urban environmental applications of GIScience: Challenges and new trends. In *Sustainable practices: Concepts, methodologies, tools and applications* (pp. 602–620). Hershey, PA: IGI Global. doi:10.4018/978-1-4666-4852-4.ch034

Oztaysi, B., Isik, M., & Ercan, S. (2013). Multi-criteria decision aid for sustainable energy prioritization using fuzzy axiomatic design. *International Journal of Energy Optimization and Engineering*, 2(1), 1–20. doi:10.4018/ijeoe.2013010101

Palantzas, G., Naniopoulos, A., & Koutitas, C. (2014). Management of environmental issues in port activities: The Hellenic caste study. *International Journal of Information Systems and Supply Chain Management*, 7(1), 40–55. doi:10.4018/ijisscm.2014010103

Pang, L., & Zhao, J. (2013). An empirical study on China's regional carbon emissions of agriculture. *International Journal of Asian Business and Information Management*, 4(4), 67–77. doi:10.4018/ijabim.2013100105

Papajorgji, P., Pinet, F., Miralles, A., Jallas, E., & Pardalos, P. (2012). Modeling: A central activity for flexible information systems development in agriculture and environment. In P. Papajorgji & F. Pinet (Eds.), *New technologies for constructing complex agricultural and environmental systems* (pp. 286–310). Hershey, PA: IGI Global. doi:10.4018/978-1-4666-0333-2.ch016

Papaspyropoulos, K. G., Christodoulou, A. S., Blioumis, V., Skordas, K. E., & Birtsas, P. K. (2011). The improvement of environmental performance in the nonprofit sector through informatics. In Z. Andreopoulou, B. Manos, N. Polman, & D. Viaggi (Eds.), *Agricultural and environmental informatics, governance and management: Emerging research applications* (pp. 359–376). Hershey, PA: IGI Global. doi:10.4018/978-1-60960-621-3.ch019

Pappis, C. P. (2011). Frameworks of policy making under climate change. In C. Pappis (Ed.), *Climate change, supply chain management and enterprise adaptation: Implications of global warming on the economy* (pp. 271–308). Hershey, PA: IGI Global. doi:10.4018/978-1-61692-800-1.ch009

Paquette, S. (2011). Applying knowledge management in the environmental and climate change sciences. In D. Schwartz & D. Te'eni (Eds.), *Encyclopedia of knowledge management* (2nd ed.; pp. 20–26). Hershey, PA: IGI Global. doi:10.4018/978-1-59904-931-1.ch003

Pechanec, V., & Vávra, A. (2013). Education portal on climate change with web GIS client. *Journal of Cases on Information Technology, 15*(1), 51–68. doi:10.4018/jcit.2013010104

Perl-Vorbach, E. (2012). Communicating environmental information on a company and inter-organizational level. In *Regional development: Concepts, methodologies, tools, and applications* (pp. 914–932). Hershey, PA: IGI Global. doi:10.4018/978-1-4666-0882-5.ch505

Perry, J., Paas, L., Arreola, M. E., Santer, E., Sharma, N., & Bellali, J. (2011). Promoting e-governance through capacity development for the global environment. In *Green technologies: Concepts, methodologies, tools and applications* (pp. 980–1010). Hershey, PA: IGI Global. doi:10.4018/978-1-60960-472-1.ch423

Pessoa, M., Fernandes, E., Nascimento de Queiroz, S., Ferracini, V., Gomes, M., & Dornelas de Souza, M. (2011). Mathematical-modelling simulation applied to help in the decision-making process on environmental impact assessment of agriculture. In H. do Prado, A. Barreto Luiz, & H. Filho (Eds.), *Computational methods for agricultural research: Advances and applications* (pp. 199–233). Hershey, PA: IGI Global. doi:10.4018/978-1-61692-871-1.ch011

Peters, E. J. (2013). Promoting rainwater harvesting (RWH) in small island developing states (SIDS): A case in the Grenadines. In H. Muga & K. Thomas (Eds.), *Cases on the diffusion and adoption of sustainable development practices* (pp. 403–438). Hershey, PA: IGI Global. doi:10.4018/978-1-4666-2842-7.ch015

Peters, E. J. (2014). Promoting rainwater harvesting (RWH) in small island developing states (SIDS): A case in the Grenadines. In Sustainable practices: Concepts, methodologies, tools and applications (pp. 1657-1682). Hershey, PA: IGI Global. doi:10.4018/978-1-4666-4852-4.ch092

Ploberger, C. (2011). A critical assessment of environmental degeneration and climate change: A multidimensional (political, economic, social) challenge for China's future economic development. *International Journal of Applied Logistics*, *2*(2), 1–16. doi:10.4018/jal.2011040101

Ploberger, C. (2013). A critical assessment of environmental degeneration and climate change: A multidimensional (political, economic, social) challenge for China's future economic development. In Z. Luo (Ed.), *Technological solutions for modern logistics and supply chain management* (pp. 212–229). Hershey, PA: IGI Global. doi:10.4018/978-1-4666-2773-4.ch014

Ploberger, C. (2013). China's environmental issues, a domestic challenge with regional and international implications. *International Journal of Applied Logistics*, *4*(3), 47–61. doi:10.4018/jal.2013070104

Polat, E. (2012). An approach for land-use suitability assessment using decision support systems, AHP and GIS. In O. Ercoskun (Ed.), *Green and ecological technologies for urban planning: Creating smart cities* (pp. 212–233). Hershey, PA: IGI Global. doi:10.4018/978-1-61350-453-6.ch012

Polat, E. (2013). An approach for land-use suitability assessment using decision support systems, AHP and GIS. In *Data mining: Concepts, methodologies, tools, and applications* (pp. 2153–2173). Hershey, PA: IGI Global. doi:10.4018/978-1-4666-2455-9.ch110

Pülzl, H., & Wydra, D. (2013). The evaluation of the implementation of sustainability norms: An exercise for experts or citizens? In E. Carayannis (Ed.), *Creating a sustainable ecology using technology-driven solutions* (pp. 32–45). Hershey, PA: IGI Global. doi:10.4018/978-1-4666-3613-2.ch003

Pusceddu, C. (2012). Grenelle environment project: An institutional tool for building collaborative environmental policies at a national level. In M. Tortora (Ed.), *Sustainable systems and energy management at the regional level: Comparative approaches* (pp. 348–364). Hershey, PA: IGI Global. doi:10.4018/978-1-61350-344-7.ch018

Puškaric, A., Subic, J., & Bekic, B. (2013). Regionalization as a factor of agriculture development of the Republic of Serbia. *International Journal of Sustainable Economies Management*, *2*(1), 46–54. doi:10.4018/ijsem.2013010105

Rafferty, J. M. (2012). Design of outdoor and environmentally integrated learning spaces. In M. Keppell, K. Souter, & M. Riddle (Eds.), *Physical and virtual learning spaces in higher education: Concepts for the modern learning environment* (pp. 51–70). Hershey, PA: IGI Global. doi:10.4018/978-1-60960-114-0.ch004

Rahim, R. E., & Rahman, A. A. (2014). Green IT capability and firm's competitive advantage. *International Journal of Innovation in the Digital Economy*, *5*(1), 41–49. doi:10.4018/ijide.2014010104

Raj, P. P., & Azeez, P. A. (2012). Public on conserving an urban wetland: A case from Kerala, India. In E. Carayannis (Ed.), *Sustainable policy applications for social ecology and development* (pp. 1–7). Hershey, PA: IGI Global. doi:10.4018/978-1-4666-1586-1.ch001

Rasulev, B., Leszczynska, D., & Leszczynski, J. (2014). Nanoparticles: Towards predicting their toxicity and physico-chemical properties. In Nanotechnology: Concepts, methodologies, tools, and applications (pp. 1071-1089). Hershey, PA: IGI Global. doi:10.4018/978-1-4666-5125-8.ch049

Rene, E. R., Behera, S. K., & Park, H. S. (2012). Predicting adsorption behavior in engineered floodplain filtration system using backpropagation neural networks. In S. Kulkarni (Ed.), *Machine learning algorithms for problem solving in computational applications: Intelligent techniques* (pp. 179–194). Hershey, PA: IGI Global. doi:10.4018/978-1-4666-1833-6.ch011

Rene, E. R., López, M. E., Park, H. S., Murthy, D. V., & Swaminathan, T. (2012). ANNs for identifying shock loads in continuously operated biofilters: Application to biological waste gas treatment. In M. Khan & A. Ansari (Eds.), *Handbook of research on industrial informatics and manufacturing intelligence: Innovations and solutions* (pp. 72–103). Hershey, PA: IGI Global. doi:10.4018/978-1-4666-0294-6.ch004

Rene, E. R., López, M. E., Veiga, M. C., & Kennes, C. (2011). Artificial neural network modelling for waste: Gas and wastewater treatment applications. In B. Igelnik (Ed.), *Computational modeling and simulation of intellect: Current state and future perspectives* (pp. 224–263). Hershey, PA: IGI Global. doi:10.4018/978-1-60960-551-3.ch010

Rivas, A. A., Kahn, J. R., Freitas, C. E., Hurd, L. E., & Cooper, G. (2013). The role of payments for ecological services in the sustainable development and environmental preservation of the rainforest: A case study of Barcelos, Amazonas, BR. *International Journal of Social Ecology and Sustainable Development*, *4*(3), 13–27. doi:10.4018/jsesd.2013070102

Rodrigues dos Anjos, M., & Schulz, M. (2013). Investigation of deforestation of environmental protection areas of Madeira River permanent preservation areas in Rondônia Amazon, Brazil. In E. Carayannis (Ed.), *Creating a sustainable ecology using technology-driven solutions* (pp. 335–343). Hershey, PA: IGI Global. doi:10.4018/978-1-4666-3613-2.ch023

Rojas-Mora, J., Josselin, D., Aryal, J., Mangiavillano, A., & Ellerkamp, P. (2013). The weighted fuzzy barycenter: Definition and application to forest fire control in the PACA region. *International Journal of Agricultural and Environmental Information Systems, 4*(4), 48–67. doi:10.4018/ijaeis.2013100103

Rolim da Paz, A., Uvo, C., Bravo, J., Collischonn, W., & Ribeiro da Rocha, H. (2011). Seasonal precipitation forecast based on artificial neural networks. In H. do Prado, A. Barreto Luiz, & H. Filho (Eds.), *Computational methods for agricultural research: Advances and applications* (pp. 326–354). Hershey, PA: IGI Global. doi:10.4018/978-1-61692-871-1.ch016

Romano, B., & Zullo, F. (2013). Models of urban land use in Europe: Assessment tools and criticalities. *International Journal of Agricultural and Environmental Information Systems, 4*(3), 80–97. doi:10.4018/ijaeis.2013070105

Rosen, M., Krichevsky, T., & Sharma, H. (2011). Strategies for a sustainable enterprise. In B. Unhelkar (Ed.), *Handbook of research on green ICT: Technology, business and social perspectives* (pp. 1–28). Hershey, PA: IGI Global. doi:10.4018/978-1-61692-834-6.ch001

Roussey, C., Pinet, F., & Schneider, M. (2013). Representations of topological relations between simple regions in description logics: From formalization to consistency checking. *International Journal of Agricultural and Environmental Information Systems, 4*(2), 50–69. doi:10.4018/jaeis.2013040105

Rushforth, R., & Phillips, C. F. (2012). Gathering under a green umbrella: collaborative rainwater harvesting at the University of Arizona. In E. Carayannis (Ed.), *Sustainable policy applications for social ecology and development* (pp. 139–149). Hershey, PA: IGI Global. doi:10.4018/978-1-4666-1586-1.ch011

Ruutu, J., Nurminen, J. K., & Rissanen, K. (2013). Energy efficiency of mobile device recharging. *International Journal of Handheld Computing Research, 4*(1), 59–69. doi:10.4018/jhcr.2013010104

Saïdi, S., Camara, A., Gazull, L., Passouant, M., & Soumaré, M. (2013). Lowlands mapping in forest Guinea. *International Journal of Agricultural and Environmental Information Systems, 4*(1), 20–34. doi:10.4018/jaeis.2013010102

Salewicz, K. A., Nakayama, M., & Bruch, C. (2011). Building capacity for better water decision making through internet-based decision support systems. In *Green technologies: Concepts, methodologies, tools and applications* (pp. 466–492). Hershey, PA: IGI Global. doi:10.4018/978-1-60960-472-1.ch306

Salter, S. J. (2011). When low-carbon means low-cost: Putting lessons from nature to work in our cities. *International Journal of Social Ecology and Sustainable Development*, 2(4), 12–25. doi:10.4018/jsesd.2011100102

Salter, S. J. (2013). When low-carbon means low-cost: putting lessons from nature to work in our cities. In E. Carayannis (Ed.), *Creating a sustainable ecology using technology-driven solutions* (pp. 282–295). Hershey, PA: IGI Global. doi:10.4018/978-1-4666-3613-2.ch019

Salvadó, J. A., López, J. E., & Martín de Castro, G. (2012). Social innovation, environmental innovation, and their effect on competitive advantage and firm performance. In F. Nobre, D. Walker, & R. Harris (Eds.), *Technological, managerial and organizational core competencies: Dynamic innovation and sustainable development* (pp. 89–104). Hershey, PA: IGI Global. doi:10.4018/978-1-61350-165-8.ch006

Saroar, M. M., & Routray, J. K. (2013). Desert in Bengal Delta- Changes in landscape, changes in livelihood: Can diffusion and adoption of sustainable adaptation make a difference? In H. Muga & K. Thomas (Eds.), *Cases on the diffusion and adoption of sustainable development practices* (pp. 83–117). Hershey, PA: IGI Global. doi:10.4018/978-1-4666-2842-7.ch004

Saroar, M. M., & Routray, J. K. (2014). Desert in Bengal Delta-Changes in landscape, changes in livelihood: Can diffusion and adoption of sustainable adaptation make a difference? In *Sustainable practices: Concepts, methodologies, tools and applications* (pp. 1414–1441). Hershey, PA: IGI Global. doi:10.4018/978-1-4666-4852-4.ch079

Schmehl, M., Eigner-Thiel, S., Ibendorf, J., Hesse, M., & Geldermann, J. (2012). Development of an information system for the assessment of different bioenergy concepts regarding sustainable development. In *Regional development: Concepts, methodologies, tools, and applications* (pp. 274–292). Hershey, PA: IGI Global. doi:10.4018/978-1-4666-0882-5.ch206

Schröter, M., Jakoby, O., Olbrich, R., Eichhorn, M., & Baumgärtner, S. (2011). Remote sensing of bush encroachment on commercial cattle farms in semi-arid rangelands in Namibia. In V. Olej, I. Obršálová, & J. Krupka (Eds.), *Environmental modeling for sustainable regional development: System approaches and advanced methods* (pp. 327–343). Hershey, PA: IGI Global. doi:10.4018/978-1-60960-156-0.ch016

Schultz, R. A. (2014). Environmentalism and sustainability. In *Technology versus ecology: Human superiority and the ongoing conflict with nature* (pp. 180–212). Hershey, PA: IGI Global. doi:10.4018/978-1-4666-4586-8.ch010

Schultz, R. A. (2014). More about modern technology. In *Technology versus ecology: Human superiority and the ongoing conflict with nature* (pp. 145–158). Hershey, PA: IGI Global. doi:10.4018/978-1-4666-4586-8.ch008

Schultz, R. A. (2014). The role of science and technology. In *Technology versus ecology: Human superiority and the ongoing conflict with nature* (pp. 213–230). Hershey, PA: IGI Global. doi:10.4018/978-1-4666-4586-8.ch011

Sdrali, D., Galanis, N., Goussia-Rizou, M., & Abeliotis, K. (2014). Are Greek high school students environmental citizens? A cluster analysis approach. *International Journal of Information Systems and Social Change*, 5(1), 16–29. doi:10.4018/ijissc.2014010102

Selmaoui-Folcher, N., Flouvat, F., Gay, D., & Rouet, I. (2012). Spatial pattern mining for soil erosion characterization. In P. Papajorgji & F. Pinet (Eds.), *New technologies for constructing complex agricultural and environmental systems* (pp. 190–210). Hershey, PA: IGI Global. doi:10.4018/978-1-4666-0333-2.ch011

Shahid, M., Mishra, H., Mishra, H. K., Tripathi, T., Khan, H. M., Sobia, F., & Singh, A. (2012). Pharmaco-ecomicrobiology and its potential role in medical and environmental sciences. In T. Gasmelseid (Ed.), *Pharmacoinformatics and drug discovery technologies: Theories and applications* (pp. 291–302). Hershey, PA: IGI Global. doi:10.4018/978-1-4666-0309-7.ch018

Shakir, I., Ali, Z., Rana, U. A., Nafady, A., Sarfraz, M., Al-Nashef, I., ... Kang, D. (2014). Nanostructured materials for the realization of electrochemical energy storage and conversion devices: Status and prospects. In M. Bououdina & J. Davim (Eds.), *Handbook of research on nanoscience, nanotechnology, and advanced materials* (pp. 376–413). Hershey, PA: IGI Global. doi:10.4018/978-1-4666-5824-0.ch015

Sharma, P., Hussain, N., Das, M. R., Deshmukh, A. B., Shelke, M. V., Szunerits, S., & Boukherroub, R. (2014). Metal oxide-graphene nanocomposites: Synthesis to applications. In M. Bououdina & J. Davim (Eds.), *Handbook of research on nanoscience, nanotechnology, and advanced materials* (pp. 196–225). Hershey, PA: IGI Global. doi:10.4018/978-1-4666-5824-0.ch010

Sidorov, E., & Ritschelová, I. (2011). Economic performance and environmental quality at the regional level an approach to modeling depletion adjusted macro aggregates for the Czech coal mining regions. In V. Olej, I. Obršálová, & J. Krupka (Eds.), *Environmental modeling for sustainable regional development: System approaches and advanced methods* (pp. 281–302). Hershey, PA: IGI Global. doi:10.4018/978-1-60960-156-0.ch014

Silva, M. R., & McLellan, S. L. (2012). Environmental and social impact of stormwater outfalls at Lake Michigan beaches. In E. Carayannis (Ed.), *Sustainable policy applications for social ecology and development* (pp. 150–165). Hershey, PA: IGI Global. doi:10.4018/978-1-4666-1586-1.ch012

Snyder, A. (2014). Environmental protection agency. In J. Krueger (Ed.), *Cases on electronic records and resource management implementation in diverse environments* (pp. 363–377). Hershey, PA: IGI Global. doi:10.4018/978-1-4666-4466-3.ch022

Somavat, P., & Namboodiri, V. (2012). Information and communication technology revolution and global warming. In W. Hu & N. Kaabouch (Eds.), *Sustainable ICTs and management systems for green computing* (pp. 23–44). Hershey, PA: IGI Global. doi:10.4018/978-1-4666-1839-8.ch002

Spanu, V., & McCall, M. K. (2013). Eliciting local spatial knowledge for community-based disaster risk management: Working with cybertracker in Georgian caucasus. *International Journal of E-Planning Research*, *2*(2), 45–59. doi:10.4018/ijepr.2013040104

Stewart, A. W. (2014). Integrating sustainability within higher education. In K. Thomas & H. Muga (Eds.), *Handbook of research on pedagogical innovations for sustainable development* (pp. 369–382). Hershey, PA: IGI Global. doi:10.4018/978-1-4666-5856-1.ch017

Stewart, C. O., & Rhodes, C. (2014). Global warming as a socioscientific controversy. In R. Hart (Ed.), *Communication and language analysis in the public sphere* (pp. 276–289). Hershey, PA: IGI Global. doi:10.4018/978-1-4666-5003-9.ch016

Suaire, R., Durickovic, I., Simonnot, M., & Marchetti, M. (2013). Monitoring of road deicers in a retention pond. *International Journal of Measurement Technologies and Instrumentation Engineering*, *3*(1), 39–47. doi:10.4018/ijmtie.2013010104

Subic, J., & Jelocnik, M. (2013). Economic and environmental aspects of controlled vegetable production within the region of Danube basin. In A. Jean-Vasile, T. Adrian, J. Subic, & D. Dusmanescu (Eds.), *Sustainable technologies, policies, and constraints in the green economy* (pp. 39–62). Hershey, PA: IGI Global. doi:10.4018/978-1-4666-4098-6.ch003

Tabrizi, A., & Sanguinetti, P. (2013). Case study: Evaluation of renewable energy strategies using building information modeling and energy simulation. *International Journal of 3-D Information Modeling*, *2*(4), 25-37. doi:10.4018/ij3dim.2013100103

Taylor, R., Barron, E., & Eames, K. A. (2014). Embedding sustainability learning: Robustness in changing circumstances - Perspectives from a United Kingdom (UK) higher education institution (HEI). In K. Thomas & H. Muga (Eds.), *Handbook of research on pedagogical innovations for sustainable development* (pp. 641–671). Hershey, PA: IGI Global. doi:10.4018/978-1-4666-5856-1.ch033

Thiell, M., & Zuluaga, J. P. (2013). Is it feasible to implement green logistics in emerging markets? *International Journal of Applied Logistics*, 4(1), 1–13. doi:10.4018/jal.2013010101

Thongmak, M. (2013). A systematic framework for sustainable ICTs in developing countries. *International Journal of Information Technologies and Systems Approach*, 6(1), 1–19. doi:10.4018/jitsa.2013010101

Torrens, F., & Castellano, G. (2012). Cluster origin of solvent features of fullerenes, single-wall carbon nanotubes, nanocones, and nanohorns. In E. Castro & A. Haghi (Eds.), *Nanoscience and advancing computational methods in chemistry: Research progress* (pp. 1–57). Hershey, PA: IGI Global. doi:10.4018/978-1-4666-1607-3.ch001

Torrens, F., & Castellano, G. (2014). Cluster origin of solvent features of fullerenes, single-wall carbon nanotubes, nanocones, and nanohorns. In I. Management Association (Ed.), Nanotechnology: Concepts, methodologies, tools, and applications (pp. 262-318). Hershey, PA: IGI Global. doi:10.4018/978-1-4666-5125-8.ch011

Touza, L. L., & López-Gunn, E. (2012). Climate change policies—Mitigation and adaptation at the local level: The case of the city of Madrid (Spain). In M. Tortora (Ed.), *Sustainable systems and energy management at the regional level: Comparative approaches* (pp. 261–287). Hershey, PA: IGI Global. doi:10.4018/978-1-61350-344-7.ch014

Trautmann, N. M., & McLinn, C. M. (2012). Using online data for student investigations in biology and ecology. In A. Olofsson & J. Lindberg (Eds.), *Informed design of educational technologies in higher education: Enhanced learning and teaching* (pp. 80–100). Hershey, PA: IGI Global. doi:10.4018/978-1-61350-080-4.ch005

Trisurat, Y., Shrestha, R. P., & Alkemade, R. (2011). Linkage between biodiversity, land use informatics and climate change. In Y. Trisurat, R. Shrestha, & R. Alkemade (Eds.), *Land use, climate change and biodiversity modeling: Perspectives and applications* (pp. 1–22). Hershey, PA: IGI Global. doi:10.4018/978-1-60960-619-0.ch001

Trivedi, B. (2011). Developing environmentally responsible business strategies: A research perspective. *International Journal of Green Computing*, 2(1), 47–57. doi:10.4018/jgc.2011010105

Trivedi, B. (2013). Developing environmentally responsible business strategies: A research perspective. In K. Ganesh & S. Anbuudayasankar (Eds.), *International and interdisciplinary studies in green computing* (pp. 47–57). Hershey, PA: IGI Global. doi:10.4018/978-1-4666-2646-1.ch005

Tsalapata, H., Alimsi, R., & Heidmann, O. (2012). Environmental education through envkids didactical framework and ICT tools. In Z. Lu (Ed.), *Learning with mobile technologies, handheld devices, and smart phones: Innovative methods* (pp. 147–161). Hershey, PA: IGI Global. doi:10.4018/978-1-4666-0936-5.ch009

Tsalapata, H., Alimsi, R., & Heidmann, O. (2014). Environmental education through envkids didactical framework and ICT tools. In *Sustainable practices: Concepts, methodologies, tools and applications* (pp. 1492–1504). Hershey, PA: IGI Global. doi:10.4018/978-1-4666-4852-4.ch083

Turek, A. (2013). Sustainable agriculture: between sustainable development and economic competitiveness. In A. Jean-Vasile, T. Adrian, J. Subic, & D. Dusmanescu (Eds.), *Sustainable technologies, policies, and constraints in the green economy* (pp. 219–235). Hershey, PA: IGI Global. doi:10.4018/978-1-4666-4098-6.ch012

Turgut, E. T., & Rosen, M. A. (2012). Emission assessment of aviation. In E. Abu-Taieh, A. El Sheikh, & M. Jafari (Eds.), *Technology engineering and management in aviation: Advancements and discoveries* (pp. 20–72). Hershey, PA: IGI Global. doi:10.4018/978-1-60960-887-3.ch002

Twesigye, C. K. (2011). Application of remote sensing technologies and geographical information systems in monitoring environmental degradation in the Lake Victoria watershed, East Africa. In *Green technologies: Concepts, methodologies, tools and applications* (pp. 653–677). Hershey, PA: IGI Global. doi:10.4018/978-1-60960-472-1.ch405

Uchida, S., Hayashi, K., Sato, M., & Hokazono, S. (2011). Construction of agri-environmental data using computational methods: The case of life cycle inventories for agricultural production systems. In H. do Prado, A. Barreto Luiz, & H. Filho (Eds.), *Computational methods for agricultural research: advances and applications* (pp. 412–433). Hershey, PA: IGI Global. doi:10.4018/978-1-61692-871-1.ch019

Unhelkar, B. (2011). Green enterprise architecture using environmental intelligence. *International Journal of Green Computing*, 2(1), 58–65. doi:10.4018/jgc.2011010106

Unhelkar, B., & Trivedi, B. (2011). A framework for environmentally responsible business strategies. In B. Unhelkar (Ed.), *Handbook of research on green ICT: Technology, business and social perspectives* (pp. 214–232). Hershey, PA: IGI Global. doi:10.4018/978-1-61692-834-6.ch014

Urban, M. J., Marker, E., & Falvo, D. A. (2012). An interdisciplinary exploration of the climate change issue and implications for teaching STEM through inquiry. In L. Lennex & K. Nettleton (Eds.), *Cases on inquiry through instructional technology in math and science* (pp. 523–550). Hershey, PA: IGI Global. doi:10.4018/978-1-4666-0068-3.ch019

Urooj, S., Hussain, A., & Srivastava, N. (2013). Biodiesel production from algal blooms: A step towards renewable energy generation & measurement. *International Journal of Measurement Technologies and Instrumentation Engineering*, 2(3), 60–71. doi:10.4018/ijmtie.2012070106

Uyttersprot, I., & Vermeir, I. (2014). Should I recycle or not? Effects of attitude strength and social pressure. In A. Kapoor & C. Kulshrestha (Eds.), *Dynamics of competitive advantage and consumer perception in social marketing* (pp. 292–308). Hershey, PA: IGI Global. doi:10.4018/978-1-4666-4430-4.ch012

Varga, L., Camci, F., Boxall, J., Toossi, A., Machell, J., Blythe, P. T., & Taylor, C. (2013). Transforming critical infrastructure: Matching the complexity of the environment to policy. *International Journal of E-Planning Research*, 2(3), 38–49. doi:10.4018/ijepr.2013070104

Viaggi, D., & Raggi, M. (2011). Estimation of irrigation water demand on a regional scale combining positive mathematical programming and cluster analysis in model calibration. In V. Olej, I. Obršálová, & J. Krupka (Eds.), *Environmental modeling for sustainable regional development: System approaches and advanced methods* (pp. 204–220). Hershey, PA: IGI Global. doi:10.4018/978-1-60960-156-0.ch010

Wang, H. (2011). GHG emissions from the international goods movement by ships and the adaptation funding distribution. In Z. Luo (Ed.), *Green finance and sustainability: Environmentally-aware business models and technologies* (pp. 274–290). Hershey, PA: IGI Global. doi:10.4018/978-1-60960-531-5.ch015

Wang, H., & Ghose, A. K. (2011). Green strategic alignment: Aligning business strategies with sustainability objectives. In B. Unhelkar (Ed.), *Handbook of research on green ICT: Technology, business and social perspectives* (pp. 29–41). Hershey, PA: IGI Global. doi:10.4018/978-1-61692-834-6.ch002

Wang, S., Ku, C., & Chu, C. (2013). Sustainable campus project: Potential for energy conservation and carbon reduction education in Taiwan. *International Journal of Technology and Human Interaction*, 8(3), 19–30. doi:10.4018/jthi.2012070103

Wang, Y. (2014). Texted environmental campaign in China: A case study of new media communication. In J. Pelet & P. Papadopoulou (Eds.), *User behavior in ubiquitous online environments* (pp. 19–43). Hershey, PA: IGI Global. doi:10.4018/978-1-4666-4566-0.ch002

Wani, A. H., Amin, M., Shahnaz, M., & Shah, M. A. (2012). Antimycotic activity of nanoparticles of MgO, FeO and ZnO on some pathogenic fungi. *International Journal of Manufacturing, Materials, and Mechanical Engineering*, 2(4), 59–70. doi:10.4018/ijmmme.2012100105

Wani, A. H., Amin, M., Shahnaz, M., & Shah, M. A. (2014). Antimycotic activity of nanoparticles of MgO, FeO and ZnO on some pathogenic fungi. In Nanotechnology: Concepts, methodologies, tools, and applications (pp. 1289-1299). Hershey, PA: IGI Global. doi:10.4018/978-1-4666-5125-8.ch059

Wati, Y., & Koo, C. (2011). A new recommendation for green IT strategies: A resource-based perspective. In Z. Luo (Ed.), *Green finance and sustainability: Environmentally-aware business models and technologies* (pp. 153–175). Hershey, PA: IGI Global. doi:10.4018/978-1-60960-531-5.ch008

Williamson, T. B., Hauer, G. K., & Luckert, M. K. (2011). Economic concepts, methods, and tools for risk analysis in forestry under climate change. In V. Olej, I. Obršálová, & J. Krupka (Eds.), *Environmental modeling for sustainable regional development: System approaches and advanced methods* (pp. 303–326). Hershey, PA: IGI Global. doi:10.4018/978-1-60960-156-0.ch015

Wong, Y. M. (2014). Fair share of supply chain responsibility for low carbon manufacturing. In Z. Luo (Ed.), *Smart manufacturing innovation and transformation: interconnection and intelligence* (pp. 303–332). Hershey, PA: IGI Global. doi:10.4018/978-1-4666-5836-3.ch012

Wu, J., & Haasis, H. (2013). Integration of knowledge management approach to the planning stage of freight villages: Towards sustainable development. *International Journal of Applied Logistics*, 4(2), 46–65. doi:10.4018/jal.2013040104

Yamaguchi, M., & Fara, L. (2013). New trends in solar cells. In L. Fara & M. Yamaguchi (Eds.), *Advanced solar cell materials, technology, modeling, and simulation* (pp. 1–21). Hershey, PA: IGI Global. doi:10.4018/978-1-4666-1927-2.ch001

Yi, L. (2011). E-business/ICT and carbon emissions. In O. Bak & N. Stair (Eds.), *Impact of e-business technologies on public and private organizations: Industry comparisons and perspectives* (pp. 214–232). Hershey, PA: IGI Global. doi:10.4018/978-1-60960-501-8.ch013

Yi, L. (2013). E-business/ICT and carbon emissions. In *Industrial engineering: Concepts, methodologies, tools, and applications* (pp. 1833–1852). Hershey, PA: IGI Global. doi:10.4018/978-1-4666-1945-6.ch098

Younessi, D. (2011). Sustainable business value. In B. Unhelkar (Ed.), *Handbook of research on green ICT: Technology, business and social perspectives* (pp. 98–115). Hershey, PA: IGI Global. doi:10.4018/978-1-61692-834-6.ch007

Yu, T., Lenzen, M., & Dey, C. (2011). Large-scale computational modeling for environmental impact assessment. In V. Olej, I. Obršálová, & J. Krupka (Eds.), *Environmental modeling for sustainable regional development: system approaches and advanced methods* (pp. 1–17). Hershey, PA: IGI Global. doi:10.4018/978-1-60960-156-0.ch001

Zhu, Y., & Heath, T. (2012). Technologies in urban design practice: Integrating environmental design considerations. In O. Ercoskun (Ed.), *Green and ecological technologies for urban planning: Creating smart cities* (pp. 133–152). Hershey, PA: IGI Global. doi:10.4018/978-1-61350-453-6.ch008

Zoltáni, T. A. (2013). Carbon as an emerging tool for risk management. *International Journal of Applied Logistics*, *4*(4), 51–69. doi:10.4018/ijal.2013100104

Compilation of References

"Science for Environment Policy": European Commission DG Environment News Alert Service. (2016, May 6). SCU, The University of the West of England.

Abbasi, T., & Abbasi, S. A. (2012). *Water Quality Indices*. Amsterdam: Elsevier. doi:10.1016/B978-0-444-54304-2.00016-6

Adams, P. (2015). The truth about China; Why Beijing will resist demands for abatement. *The Global Warming Policy Foundation GWPF Report 19*.

Adams, A. (1993). Food Insecurity in Mali: Exploring the Role of the Moral Economy. *IDA Bulletin*, *24*(4), 41–51. doi:10.1111/j.1759-5436.1993.mp24004005.x

Adams, P. (2015). *The truth about China; Why Beijing will resist demands for abatement*. The Global Warming Policy Foundation.

AirNow. (n.d.). *Current AQUI*. Retrieved from https://www.airnow.gov

Alam, U. (2013). International Law and Freshwater: The Multiple Challenges. Cheltenhem, UK: Edward Elgar Publishing.

Albayrak, T., Caber, M., Moutinho, L., & Herstein, R. (2011). The influence of skepticism on green purchase behavior. *International Journal of Business and Social Science*, *2*(13), 189–197.

Amadi, L & Agena, J .(2015).Globalization Culture Mutation and New Identity. *African Journal of Culture and History*, *7*(1), 16-27.

Amadi, L., & Anokwuru, G. (2017). Sustainable Rural Livelihoods: Elusive Post-Colonial Development Project in Nigeria? *International Journal of Poultry Science*, *2*(4), 1–16.

Amadi, L., & Igwe, P. (2016). Maximizing the Eco Tourism Potentials of the Wetland Regions through Sustainable Environmental Consumption: A Case of the Niger Delta, Nigeria. *The Journal of Social Sciences Research*, *2*(1), 13–22.

Amadi, L., Igwe, P., & Ogbanga, M. (2016). Talking Right, Walking Wrong: Global Environmental Negotiations and Unsustainable Environmental Consumption. *International Journal of Research in Environmental Science*, *2*(2), 24–38.

Compilation of References

Amadi, L., & Imoh-ita, I. (2017). Intellectual capital and environmental sustainability measurement nexus: A review of the literature *Int. J. Learning and Intellectual Capital, X14*(2), 154–176. doi:10.1504/IJLIC.2017.084071

Amadi, L., Wordu, S., & Ogbanga, M. (2015). Sustainable Development in Crisis? A Post Development Perspective. *Journal of Sustainable Development in Africa, 17*(1), 140–163.

Amburgey, J. W., & Thoman, D. B. (2012). Dimensionality of the New Ecological Paradigm Issues of Factor Structure and Measurement. *Environment and Behavior, 44*(2), 235–256. doi:10.1177/0013916511402064

Anders, E. (2008). *China environment and climate change policy brief.* University of Gothenburg.

Anderson, B. A., Romani, J. H., Phillips, H., Wentzel, M., & Tlabela, K. (2007). Exploring environmental perceptions, behaviors and awareness: Water and water pollution in South Africa. *Population and Environment, 28*(3), 133–161. doi:10.100711111-007-0038-5

Andersson, L. (2006). *China's environmental policy process: A case study of the challenges of creating and implementing air pollution policies in China.* University of Lund, Department of Political Science.

Angell, D. J., Comer, J. D., & Wilkinson, M. L. (Eds.). (1990). *Sustaining Earth: response to the environmental threat.* London: Macmillan. doi:10.1007/978-1-349-21091-6

Ansari, A. A., & Gill, S. S. (2014). *Eutrophication: causes, consequences and control* (Vol. 2). London: Springer. doi:10.1007/978-94-007-7814-6

Appadurai, A. (1996). *Modernity at Large: Cultural Dimensions of Globalization.* Minneapolis, MN: University of Minnesota Press.

Appadurai, A. (2003). Modernity at Large: Cultural Dimensions of Globalization, *The Russian. The Sociological Review, 3*(4), 57–66.

Ares, E. (2012). *Durban Climate Conference.* Science and Environment Section SN/SC/6140.

Armstrong, D., Lloyd, L., & Redmond, J. (2004). *International Organization in World Politics. Hound mills.* Basingstoke, UK: Palgrave Macmillan. doi:10.1007/978-0-230-62952-3

Artaza, I. (2017, March 12). Water Security. *Dawn.* Retrieved from: https://www.dawn.com/news/1319986

Asian Development Bank (ADB). (2012). *World Sustainable Development Timeline.* Retrieved from https://www.researchgate.net/profile/Olivier_Serrat/publication/266878643_World_Sustainable_Development_Timeline/links/543e3e430cf2d6934ebd20af/World-Sustainable-Development-Timeline.pdf?origin=publication_list

Awasthi, P. (2011). Socio-Economic Challenges and Sustainable Development in Developing Countries. *Management Insight, 7*(2), 56–63.

Baeumler, A., Ijjasz-Vasquez, E., & Mehndiratta, S. (2012). *Sustainable low-carbon city development in China.* The World Bank.

Bakan, J. (2004). *The Corporation: The Pathological Pursuit of Profit & Power.* Toronto: Penguin Group.

Barrett, S. (1998). Political Economy of the Kyoto Protocol. *Oxford Review of Economic Policy, 14*(4), 20–39. doi:10.1093/oxrep/14.4.20

Bauman, Z. (1998). *Work, Consumerism and the New Poor.* Philadelphia: Open University Press.

BBC News. (2017, August 28). *China claims victory over India in Himalayan border row.* Retrieved from: http://www.bbc.com/news/world-asia-41070767

Beck, U. (1992). *Risk Society. Towards a New Modernity.* London: Sage.

Beitz, C. (2001). Does Global Inequality Matter? In T. W. Pogge (Ed.), *Global Justice*. Oxford. doi:10.1111/1467-9973.00177

Bello, W. (2003). *De-globalization Ideas for a New World Economy.* Fernwood Publishing Limited.

Bennet, G., & Williams, F. (2011). *Mainstream Green: Moving sustainability from niche to normal.* Chicago: Ogilvy & Mather.

Berndt, A., & Petzer, D. (2011). Environmental concern of South African cohorts: An exploratory study. *African Journal of Business Management, 5*(19), 7899–7910.

Best, G. (1999). *Environmental Pollution Studies.* Liverpool: Liverpool University Press. doi:10.5949/UPO9781846313035

Bhutta, Z. (2011, November 15). Water wars: India planning 155 hydel projects on Pakistan's rivers. *The Express Tribune.* Retrieved from: http://tribune.com.pk/story/292021/water-wars-india-planning-155-hydel-projects-on-pakistans-rivers/

Bina, O. (2013). The green economy and sustainable development: An uneasy balance? *Environment and Planning. C, Government & Policy, 31*(6), 1023–1047. doi:10.1068/c1310j

Bindoff, Bindschadler, Cox, de Noblet, England, Francis, … Weaver. (2009). The Copenhagen Diagnosis: updating the world on the latest climate science. The University of New South Wales, Climate Change Research Centre (CCRC).

Birău, R. (2017) Conceptual approaches on sustainable development in public administration in Romania. *The Journal Contemporary Economy, 2*(4), 34-37.

Birău, R. (2015). Statistical analysis on poverty in European Union. A case study for Romania. *International Journal of Business Quantitative Economics and Applied Management Research, 1*(7), 1–10.

Blaikie, P., & Brookfield, H. (1987). *Land Degradation and Society.* London: Methuen.

Compilation of References

Bocchini, P., Frangopol, D., Ummenhofer, T., & Zinke, T. (2014). Resilience and Sustainability of Civil Infrastructure: Toward a Unified Approach. *Journal of Infrastructure Systems, Publisher: American Society of Civil Engineers*, *20*(2).

Bodle, R., Donat, L., & Duwe, M. (2016). *The Paris Agreement: Analysis, Assessment and Outlook. Background paper for the workshop "Beyond COP21: what does Paris mean for future climate policy?"* 28 January 2016, Federal Ministry for the Environment, Nature Conservation, Building and Nuclear Safety. Berlin: BMUB.

Bohringer, C. (2003). *The Kyoto protocol: A Review and Perspective*. ZEW discussion paper No. 03-61, Mannheim.

Bonini, S., & Oppenheim, J. (2008). Cultivating the green consumer. *Stanford Social Innovation Review*, *6*(4), 56–61.

Brennan, L., & Binney, W. (2008). Is it green marketing, greenwash or hogwash? We need to know if we want to change things. In *Proceedings of Partnerships, proof and practice: International nonprofit and social marketing conference*. University of Wollongong.

Broder, J. M. (2009a, December 19). Many Goals Remain Unmet in Five Nations Climate Deal. New York Times, pp. 2-3.

Broder, J. M. (2009b, December 17). Poor and Emerging States Stall Climate Negotiations. New York Times, pp.1-2.

Bryant, R. (1998). Power, Knowledge and Political Ecology. *Progress in Physical Geography*, *22*(1), 79–94. doi:10.1177/030913339802200104

Bump, P. (2017). Nine reasons Trump's withdrawal from the Paris climate agreement doesn't make sense Politics Analysis. *Washington Post*. Retrieved from https://www.washingtonpost.com/news/politics/wp/2017/06/01/all-the-reasons-that-trumps-withdrawal-from-the-paris-climate-agreement-doesnt-make-sense/?utm_term=.8128f4219a0d

Butler, R. (2005). *Nigeria has worst deforestation rate, FAO revises*. Available at https://news.mongabay.com/2005/11/nigeria-has-worst-deforestation-rate-fao revises-figures/Accesses 18/4/2017

Callinicos, A. (2009). *Imperialism and Global Political Economy*. London: Polity Press.

Canfa, W. (2007). Chinese environmental law enforcement: Current deficiencies and suggested reforms. *Vermont. Journal of Environmental Law*, *8*, 159–190.

Caroline, B., & Peterson, J. (2009). *Conceptualizing Multilateralism*. Mercury Working Paper.

Carson, R. (1962). *Silent Spring*. New York: Houghton Mifflin.

Castells, M. (2000a). The Information Age: Economy: Society, and Culture (2nd ed.; Vols. 1-3). Maiden, MA: Blackwell.

Castro, C. J. (2004). Sustainable Development: Mainstream and Critical Perspectives. *Organization & Environment, 17*(2), 195–225. doi:10.1177/1086026604264910

Chambers, N., Simmons, C., & Wackernagel, M. (2001). *Sharing nature's interest: Ecological footprints as an indicator of sustainability.* London: Earthscan Publications.

Chang, C. (2011). Feeling ambivalent about going green. *Journal of Advertising, 40*(4), 19–32. doi:10.2753/JOA0091-3367400402

Chan, R. (2001). Determinants of Chinese consumers' green purchase behavior. *Psychology and Marketing, 18*(4), 389–413. doi:10.1002/mar.1013

Cheever, F., & Dernbach, J. C. (2015). Sustainable Development and its Discontents. *Journal of Transnational Environmental Law.*

Chen, T. B., & Chai, L. T. (2010). Attitude towards the environment and green products: Consumers' perspective. *Management Science and Engineering, 4*(2), 27.

Chow, G. C. (2007). China's energy and environmental problems and policies. *CEPS Working Paper, 152,* 1-18.

Chow, L. (2017). *EPA Chief Denies CO2 as Primary Driver of Climate Change.* Eco Watch.

Climate focus. (2015). *The Paris Agreement: Summary.* Climate Focus Client Brief on the Paris Agreement III 28 December 2015. Briefing Note. Retrieved from http://unfccc.int/paris_agreement/items/9444.php

Collier, P. (2010). *The Plundered Planet. Why We Must--and How We Can--Manage Nature for Global Prosperity.* Oxford Press.

Collier, P., & Gunning, J. (1999). Why Has Africa Grown Slowly? *The Journal of Economic Perspectives, 13*(3), 3–22. doi:10.1257/jep.13.3.3

Cooper, A. F. (2002). Like-minded nations, NGOs, and the changing pattern of diplomacy with in the UN system: An introductory perspective. In A. F. Cooper, J. English, & R. Thakur (Eds.), *Enhancing Global Governance: Towards a New Diplomacy?* Tokyo: United Nations University Press.

D'Souza, C., Taghian, M., & Lamb, P. (2006). An empirical study on the influence of environmental labels on consumers. *Corporate Communication: An International Journal, 11*(2), 162–173. doi:10.1108/13563280610661697

Dalby, S. (2002). *Environmental Security.* Minneapolis, MN: University of Minosota Press.

Dalby, S. (2013). Environmental dimensions of human security. In R. Floyd & R. Mathew (Eds.), *Environmental Security Approaches and Issues.* London Routledge Taylor and Francis.

Dale, A., & Newman, L. (2005). Sustainable development, education and literacy. *International Journal of Sustainability in Higher Education, 6*(4), 351–362. doi:10.1108/14676370510623847

Compilation of References

Darren, S. (2009, December 21). Obama Negotiates 'Copenhagen Accord' With Senate Climate Fight in Mind. *The New York Times*, p. 1.

Dasgupta, S., & Tam, E. K. L. (2005). Indicators and framework for assessing sustainable infrastructure. *Canadian Journal of Civil Engineering*, *32*(1), 30–44. doi:10.1139/l04-101

Datta, S. K. (2011). Pro-environmental concern influencing green buying: A study on Indian consumers. *International Journal of Business and Management*, *6*(6), 124.

Davari, A., & Strutton, D. (2014). Marketing mix strategies for closing the gap between green consumers' pro-environmental beliefs and behaviors. *Journal of Strategic Marketing*, *22*(7), 563–586. doi:10.1080/0965254X.2014.914059

Davidson, D., & Hatt, K. (2005). *Consuming Sustainability Critical Social Analysis of Ecological Change*. Fernwood Publishing.

de Barcellos, M. D., Krystallis, A., de Melo Saab, M. S., Kügler, J. O., & Grunert, K. G. (2011). Investigating the gap between citizens' sustainability attitudes and food purchasing behaviour: Empirical evidence from Brazilian pork consumers. *International Journal of Consumer Studies*, *35*(4), 391–402. doi:10.1111/j.1470-6431.2010.00978.x

De Sherbinin, A., Carr, D., Cassels, S., & Jiang, L. (2007). Population and environment. *Annual Review of Environment and Resources*, *32*(1), 345–373. doi:10.1146/annurev.energy.32.041306.100243 PMID:20011237

de Soyza, I. (2000). The resource curse: Are civil wars driven by rapacity or paucity? In M. Berdal & D. M. Malone (Eds.), *Greed and grievance: Economic agendas in civil wars* (pp. 113–135). Boulder, CO: Lynne Rienner.

Derzon, J. H., & Lipsey, M. W. (2002). *A meta-analysis of the effectiveness of mass communication for changing substance use knowledge, attitudes and behavior. In Mass media and drug prevention: Classic and contemporary theories and research* (pp. 231–258). Mahwah, NJ: Erlbaum.

Doshi, V. (2016, May 18). India set to start massive project to divert Ganges and Brahmaputra rivers. *The Guardian*. Retrieved from: https://www.theguardian.com/global-development/2016/may/18/india-set-to-start-massive-project-to- divert-ganges-and-brahmaputra-rivers

Dreher, A. (2006). Does globalization affect growth? Evidence from a new index of globalization. *Applied Economics*, *38*(10), 1091–1110. doi:10.1080/00036840500392078

Dreher, A., Gaston, N., & Martens, P. (2008). *Measuring globalisation: Gauging its consequences*. New York: Springer. doi:10.1007/978-0-387-74069-0

Driessen, P. (2003). *Eco-Imperialism: Green Power, Black Death*. Belleview, WA: Free Enterprise Press.

Dunlap, R., Kent, D., Angela, G., & Robert, E. (2000). Measuring Endorsement of the New Ecological Paradigm: A revised NEP Scale. *The Journal of Social Issues*, *56*(3), 425–442. doi:10.1111/0022-4537.00176

Duruji, M., & Urenma, D.-M. (2016). The Environmentalism and Politics of Climate Change: A Study of the Process of Global Convergence through UNFCCC Conferences. In *Handbook of Research on Global Indicators of Economic and Political Convergence*. Hershey, PA: IGI Global.

El-Diraby, T., & Wang, B. (2005). E-Society Portal: Integrating Urban Highway Construction Projects into the Knowledge City. *Journal of Construction Engineering and Management*, *11*(1196), 1196-1211.

Elliot, J. A. (1999). *Sustainable Development*. London: Routledge.

Engels, S. V., Hansmann, R., & Scholz, R. W. (2010). Toward a sustainability label for food products: An analysis of experts' and consumers' acceptance. *Ecology of Food and Nutrition*, *49*(1), 30–60. doi:10.1080/03670240903433154 PMID:21883088

Environment. (n.d.a). In *English Oxford Living Dictionaries*. Retrieved from: https://en.oxforddictionaries.com/definition/environment

Environment. (n.d.b). In *Merriam-Webster Dictionary*. Retrieved from: https://www.merriam-webster.com/dictionary/environment

EPA. (2007). *Learn About Environmental Management Systems*. Retrieved July 19, 2017, from https://www.epa.gov/ems/learn-about-environmental-management-systems

Erbach, G. (2015). *Negotiating a new UN climate agreement: Challenges on the road to Paris. EU*. European Parliamentary Research Service.

Erickson, L. E., Griswold, W., Maghirang, R. G., & Urbas-zewski, B. P. (2017). Air Quality, Health and Community Action. *Journal of Environmental Protection*, *8*(10), 1057–1074. doi:10.4236/jep.2017.810067

Eskom. (2015). *Eskom IDM Programme: Focus on Housing Sector of South Africa*. Retrieved 13 May 2017 from http://www.ieadsm.org/wp/files/Ncayiyana_CapeTown.pdf

Esty, D., & Pangestu, M. (1999). *Globalization and The Environment in Asia*. United States-Asia Environmental Partnership Framing Paper.

Etzinger, A. (2011). *Eskom Intergrated demand Management*. Association of Municipal Electrical Utilities (AMEU) Convention. Retrieved 13 May 2017 from http://www.ameu.co.za/Portals/16/Conventions/Convention%202011/Papers/Eskom%20Integrated%20Demand%20Management%20-%20A%20Etzinger.pdf

Euroclima. (2012). *Key outcomes of the Durban climate change*. Retrieved from http://www.euroclima.org/en/euroclima/our-people/item/655-principales-resultados-de-la-cumbre-sobre-cambio-clim%C3%A1tico-de-durban

European Central Bank. (2017). *Growing Importance of Emerging Markets*. Retrieved 13 May 2017 from https://www.ecb.europa.eu/ecb/tasks/international/emerging/html/index.en.html

European Commission. (n.d.). *Eurostat*. Retrieved from http://ec.europa.eu/eurostat/

Compilation of References

European Union – EUROSTAT. (2015). *Sustainable development in the European Union*. Luxembourg: Publications Office of the European Union. Retrieved from http://ec.europa.eu/eurostat/documents/3217494/6975281/KS-GT-15-001-EN-N.pdf

Fahrenthold, D. A. (2009, December 19). Copenhagen Climate Talks, by the Numbers. *Washington Post*.

Falola, A. (2009, December 19). Climate Deal Falls Short of Key Goals. *Washington Post*, p. 4.

Farooq, U. (2017, November 16). *Mumbai's long shadow: What led to Hafiz Saeed's arrest*. Retrieved from: https://herald.dawn.com/news/1153708

Fayiga, A., Ipinmoroti, M., & Chirenje, T. (2017). Environmental pollution in Africa. *Environment, Development and Sustainability*. doi:10.100710668-016-9894-4

Fazil, M. D. (2017, March 8). Why India Must Refrain From a Water War With Pakistan: Threatening Pakistan's water supply will have a negative impact on India and all of South Asia. *The Diplomat*. Retrieved from: https://thediplomat.com/2017/03/why-india-must-refrain-from-a-water-war- with-pakistan/

Ferguson, J. (2006). *Decomposing Modernity: History and Hierarchy after Development*. Irvine, CA: Department of Anthropology University of California. doi:10.1215/9780822387640-008

Figge, F., & Martens, P. (2014). *Globalization Continues: The Maastricht Globalisation Index Revisited and Updated Globalizations*. Academic Press. 10.1080/14747731.2014.887389

Figge, L., Oebels, K., & Offermans, A. (2017). The effects of globalization on Ecological Footprints: an empirical analysis. *Environment, Development and Sustainability*, *19*(3), 863–876. doi:10.100710668-016-9769-8

Fisher, C., Bachman, B., & Bashyal, S. (2012). Demographic impacts on environmentally friendly purchase behaviors. *Journal of Targeting, Measurement and Analysis for Marketing*, *20*(3), 172–184. doi:10.1057/jt.2012.13

Flynn, R., Bellaby, P., & Ricci, M. (2009). The 'value-action gap' in public attitudes towards sustainable energy: The case of hydrogen energy. *The Sociological Review*, *57*(2_suppl), 159–180. doi:10.1111/j.1467-954X.2010.01891.x

Forbes. (2017). *11 Companies Considered Best For The Environment*. Retrieved July 18, 2017, from https:/ /www.forbes.com/sites/susanadams /2014/04/22/11- companies- considered- best-for-the-environment/#78b5119212ae

Frankfurt School-UNEP Centre/BNEF. (2014). *Global Trends in Renewable Energy Investment 2014*. Retrieved from http://fs-unep-centre.org/publications/gtr-2014

Freeman, C. W., & Lu, X. (2008). *Assessing Chinese government response to the challenge of environment and health*. Center for Strategic and International Studies.

Fridell, R. (2006). *Environmental Issues*. Marshallcavendish.

Frosch. (2017). *Eco-quality in 9 aspects*. Retrieved July 18, 2017, from http://www.frosch.de/Brand/Eco-quality-in-9-aspects

FTSE. (2018). *Country Classification*. Retrieved from http://www.ftse.com/products/indices/country-classification

Gan, C., Wee, H. Y., Ozanne, L., & Kao, T.-H. (2008). Consumers' purchasing behavior towards green products in New Zealand. *Innovative Marketing*, *4*(1), 93–102.

Garcia, S. M. (2003). *The ecosystem approach to fisheries: issues, terminology, principles, institutional foundations, implementation and outlook (No. 443)*. Rome: Food & Agriculture Org.

Garman, J. (2009, December 20). Copenhagen-Historic Failure That Will Live in Infamy. The Independence, pp. 3-4.

Garver, J. (2001). *Protracted Contest: Sino-Indian Rivalry in the Twentieth Century*. Seattle, WA: University of Washington Press.

George, C. (2007). Sustainable Development and Global Governance. *Journal of Environment & Development*, *16*(1), 102–125. doi:10.1177/1070496506298147

Giddens, A. (1999). *Runaway World: How Globalization is Reshaping our Lives*. London: Profile.

Gifford, R., & Nilsson, A. (2014). Personal and social factors that influence pro-environmental concern and behaviour: A review. *International Journal of Psychology*, *49*(3), 141–157. PMID:24821503

Global Ecolabelling Network. (2011). *What Is Ecolabelling?* Retrieved 13 May 2017 from http://www.globalecolabelling.net/what_is_ecolabelling

Goldemberg, J., & Lucon, O. (2010). *Energy, environment and development* (2nd ed.). London: Earthscan Publications Ltd.

GoLite. (2017). *Gear to Go the Distance*. Retrieved July 18, 2017, from http://www.golite.com/

Gollnow, S. (2016). *Better Water, Better Jobs – Envisioning a Sustainable Pakistan*. Retrieved from: https://sdpi.org/publications/files/Better-Water_Better-Jobs_Envisioning-a-Sustainable-Pakistan.pdf

Goodland, R. (1995). The Concept of Environmental Sustainability. *Annual Review of Ecology and Systematics*, *26*(1), 1–24. doi:10.1146/annurev.es.26.110195.000245

GrailResearch. (2011). *The Green Evolution*. Retrieved 13 May 2017 from http://www.grailresearch.com/pdf/Blog/Grail-Research-Green-Evolution-Study_240.pdf

Grankvist, G., Dahlstrand, U., & Biel, A. (2004). The impact of environmental labelling on consumer preference: Negative vs. positive labels. *Journal of Consumer Policy*, *27*(2), 213–230. doi:10.1023/B:COPO.0000028167.54739.94

Compilation of References

Hair, J., Bush, R., & Ortinau, D. (2009). *Marketing Research: In a digital information environment* (4th ed.). Boston: McGraw-Hill.

Haley, E., Hatt, K., & Tunstall, R. (2005). *You are What You eat. In Consuming Sustainability Critical Social Analysis of Ecological Change*. Fernwood Publishing.

Hali, S. M. (2004). *The Baglihar imbroglio*. Retrieved from: http://www.infopak.gov.pk

Hampson, F. O., & Reid, H. (2003). Coalition Diversity and Normative Legitimacy in Human Security Negotiations. *International Negotiation*, *8*(1), 7–42. doi:10.1163/138234003769590659

Han, H., Hsu, L.-T. J., & Lee, J.-S. (2009). Empirical investigation of the roles of attitudes toward green behaviors, overall image, gender, and age in hotel customers' eco-friendly decision-making process. *International Journal of Hospitality Management*, *28*(4), 519–528. doi:10.1016/j.ijhm.2009.02.004

Harris, P. G. (Ed.). (2002). *The environment, international relations, and US foreign policy*. Washington, DC: Georgetown University Press.

Hartmann, P., & Apaolaza-Ibáñez, V. (2012). Consumer attitude and purchase intention toward green energy brands: The roles of psychological benefits and environmental concern. *Journal of Business Research*, *65*(9), 1254–1263. doi:10.1016/j.jbusres.2011.11.001

Harvey, D. (2005). *Brief History of Neoliberalism*. New York: Oxford University Press.

Hatt, V. (2015, January 13). *Readers Supported News*. Retrieved from http://readersupportednews.org/opinion2/271-38/28024-global-warming-is-a-national-security-threat

Hawken, P., Lovins, A., & Lovins, L. (1999). *Natural capitalism: creating the next industrial revolution*. Boston: Little, Brown and Company.

He, D., Wu, R., Feng, Y., Li, Y., Ding, C., Wang, W., & Yu, D. W. (2014). REVIEW: China's transboundary waters: new paradigms for water and ecological security through applied ecology. *Journal of Applied Ecology*, *51*(5), 1159–1168. doi:10.1111/1365-2664.12298 PMID:25558084

Herald, D. (2005). *Wullar Barrage: Indo-Pak dialogue tomorrow*. Retrieved from: http://www.deccanherald.com/deccanherald/jun272005/national213051200 5626.asp

Hill, D. (2012). Alternative Institutional Arrangements: Managing Transboundary Water Resources in South Asia. *Harvard Asia Quarterly*, *14*(3). Retrieved from: http://southasiainstitute.harvard.edu/website/wp-content/uploads/2012/10/HAQ-14.31.pdf

Ho, S. (2017). Power Asymmetry in the China-India Brahmaputra River Dispute. *Asia Pacific Bulletin*, *371*. Retrieved from: https://www.eastwestcenter.org/system/tdf/private/apb371.pdf?file=1&type=node&id=35993

Hobson, K. (2003). Consumption, Environmental Sustainability and Human Geography in Australia: A Missing Research Agenda? *Australian Geographical Studies*, *41*(2), 148–155. doi:10.1111/1467-8470.00201

Hoekstra, A., & Wiedmann, T. (2014). Humanity's unsustainable environmental footprint. *Science*, *344*(6188), 1114-1117. DOI: 10.1126 science.1248365

Homer-Dixon, T. (1991). On the Threshold: Environmental Changes as Causes of Acute Conflict. *International Security*, *16*(2), 76–116. doi:10.2307/2539061

Honabarger, D. (2011). *Bridging the Gap: The Connection Between Environmental Awareness, Past Environmental Behavior, and Green Purchasing* (Masters thesis). American University, Washington, DC. Retrieved 13 May 2017 from http://www.american.edu/soc/communication/upload/Darcie-Honabarger.pdf

Hopwood, B., Mellor, M., & O'Brien, G. (2005). Sustainable development: Mapping different approaches. *Sustainable Development*, *13*(1), 38–52. doi:10.1002d.244

Hoskisson, R. E., Eden, L., Lau, C. M., & Wright, M. (2000). Strategy in Emerging Economies. *Academy of Management Journal*, *43*(3), 249–267. doi:10.2307/1556394

Houghton, J. (2009). *Global warming: the complete briefing* (4th ed.). Cambridge, UK: Cambridge University Press. doi:10.1017/CBO9780511841590

Howard, J., & Wheeler, J. (2015). What community development and citizen participation should contribute to the new global framework for sustainable development, *Community Development Journal*. Oxford Journals, *50*(4), 552–570.

Huang, Z. (2015). Case Study on the Water Management of Yaluzangbu/Brahmaputra River. *Georgetown International Environmental Review, 27*(2), 229. Retrieved from: https://gielr.files.wordpress.com/2015/04/huang-final-pdf-27-2.pdf

IISD. (1997). UN Climate Change Conference. *Earth Negotiation Bulletin*.

Imber, M., & Vogler, J. (Eds.). (2005). *Environment and International Relations*. London: Routledge.

Immerzeel, W. W., van Beek, L. P., & Bierkens, M. F. (2010). Climate change will affect the Asian water towers. *Science*, *328*(5984), 1382–1385. doi:10.1126cience.1183188 PMID:20538947

India-failed-to-provide-evidence-on-Pathankot-attack-Report/articleshow/51665009.cms

International Institute for Sustainable Development (IISD). (2012). *Sustainable Development Timeline*. Retrieved from https://www.iisd.org/pdf/2012/sd_timeline_2012.pdf

Iqbal, A. R. (2010). Water Shortage In Pakistan – A Crisis Around The Corner. *ISSRA Papers, 2*(2), 1-13. Retrieved from: http://www.ndu.edu.pk/issra/issra_pub/articles/issra- paper/ISSRA_Papers_Vol2_IssueII_2010/01-Water-Shortage-in-Pakistan-Abdul-Rauf-Iqbal.pdf

Jamwal, A. (2013). *River water Interests/disputes with India's Neighbours as Potential Flash Points*. Institute of Chinese Studies. Retrieved from: http://www.icsin.org/uploads/2015/04/16/647c4123483c6769eceb7b7a10eacf5a.pdf

Jang. (2017, July 30). *Sind Taas Muahida per Pakistan aur Almi Bank k darmiyan baat cheet ka dusra round kal Washington me hoga*. Academic Press.

Compilation of References

Jansson, J., Marell, A., & Nordlund, A. (2010). Green consumer behavior: Determinants of curtailment and eco-innovation adoption. *Journal of Consumer Marketing*, 27(4), 358–370. doi:10.1108/07363761011052396

Jianxue, L. (2010). *Water Security Cooperation and China-India Interactions*. Shui ziyuan anquan hezuo yu ZhongYin guanxi de hudong.

Jones, L. (1997). *Global Warming: The Science and the Politics*. The Fraser Institute.

Jorgenson, A., & Kick, E. (2003). Globalization and the environment. *Journal of World-systems Research*, 9(2), 195–205. doi:10.5195/JWSR.2003.243

Juwaheer, T. D. (2005). An emerging environmental market in Mauritius: Myth or reality? *World Review of Entrepreneurship, Management and Sustainable Development*, 1(1), 57–76. doi:10.1504/WREMSD.2005.007753

Kagan, R. (2002). Power and Weakness. *Policy Review, 113*. Retrieved 22nd September, 2011 from the website www.policyreview.org

Kamate, S. K., Agrawal, A., Chaudhary, H., Singh, K., Mishra, P., & Asawa, K. (2009). Public knowledge, attitude and behavioural changes in an Indian population during the Influenza A (H1N1) outbreak. *The Journal of Infection in Developing Countries, 4*(1), 7-14.

Kaminsky, J. A., & Javernick-Will, A. N. (2014). The Internal Social Sustainability of Sanitation Infrastructure. *Environmental Science & Technology (ES&T)*, 48(17), 10028–10035.

Kampert, P. (2001, April 17). U.S. Takes Heat; why is Bush's Stand on Global warming Treaty Upsetting Nations around the World. *Chicago Tribune,* pp. 2-4.

Kapp, W. (1997). Environmental policies and development planning in contemporary China. The Hague, The Netherlands: Academic Press.

Kaufmann, H. R., Panni, M., & Orphanidou, Y. (2012). Factors affecting consumers' green purchasing behavior: An integrated conceptual framework. *Amfiteatru Economic*, 14(31), 50–69.

Keck, M. E., & Sikkink, K. (1998). *Activists beyond Borders*. Ithaca, NY: Cornell University Press.

Keetelaar, J. C. (2007). *Transboundary Water Issues in South Asia*. Retrieved on July 20, 2017 from: https://www.google.com.pk/url?sa=t&rct=j&q=&esrc=s&source=web&cd= 2&cad=rja&uact=8&ved=0ah UKEwiAxYKjtZ_VAhXIchQKHWGADPcQFgg7MAE& url=https%3A%2F%2Fthesis.eur.nl%2Fpub%2 F4030%2FFinal%2520version%2520thes is_Jessica%2520Keetelaar_11_03_07.pdf&usg= AFQjCNGX6FHAroi7Tux_qr729P8T5E54OQ

Kegley, C. W., & Wittkopf, E. R. (1999). *World Politics: Trend and Transformation*. New York: World Publishes.

Kennedy, E. H., Beckley, T. M., McFarlane, B. L., & Nadeau, S. (2009). Why we don't" walk the talk": Understanding the environmental values/behaviour gap in Canada. *Human Ecology Review*, 16(2), 151.

Keohane, R. (2006). The Contingent Legitimacy of Multilateralism. *Garnet Working Paper, 9*(6).

Keohane, R. O. (1986). Reciprocity in International Relations. International Organization, 27, 1-27.

Keohane, R. O., & Nye, J. S. (2000a). Introduction. In Governance in a Globalizing World. Washington, DC: Brookings Institution.

Keohane, R. O., & Nye, J. S. (2000b). Power and Interdependence. New York: Addison-Wesley Longman.

Keohane, R. O. (1990). Multilateralism: An Agenda for Research. *International Journal (Toronto, Ont.), 45*(4), 731–764. doi:10.1177/002070209004500401

Khadka, N. S. (2016, December 22). Are India and Pakistan set for water wars? *BBC News*. Retrieved from: http://www.bbc.com/news/world-asia-37521897

Khadka, N. S. (2017, September 18). China and India water 'dispute' after border stand-off. *BBC News*. Retrieved from: http://www.bbc.com/news/world-asia-41303082

Khan, M. Z. (2017, January 21). India asked to stop work on Kishanganga and Ratle projects. *Dawn*. Retrieved from: https://www.dawn.com/news/1309767

Kim, Y., & Choi, S. M. (2005). Antecedents of green purchase behavior: An examination of collectivism, environmental concern, and PCE. *Advances in Consumer Research. Association for Consumer Research (U. S.), 32*, 592.

King, E. (2015, February 2). Kyoto Protocol:10 Years of the Worlds First Climate Change treaty. *Climate Home News*. Retrieved from http://www.climatechangenews.com/2015/02/16/kyoto-protocol-10--years-of-the-world-first-climate-change-treaty/

Kin, S. K., Hao, R., & Champeau, A. (2012). *Growth, Environment, and Politics; The case study of China. Conference on China*, Seattle, WA.

Klare, M. (1996, November). Redefining Security: The New Global Schisms. *Current History (New York, N.Y.)*.

Klare, M. T. (2002). *Resource Wars: The New Land scape of Global Conflict*. New York, NY: Henry Holt and Company.

Kollmuss, A., & Agyeman, J. (2002). Mind the gap: Why do people act environmentally and what are the barriers to pro-environmental behavior? *Environmental Education Research, 8*(3), 239–260. doi:10.1080/13504620220145401

Konya, I., & Ohashiz, H. (2004). *Globalization and Consumption Patterns among the OECD Countries*. Boston College Working Papers in Economics.

Korten, D. (1995). *When Corporations Rule the World*. Kumarian Press Inc./Berrett-Koehler.

Kostka, G. (2014). Barriers to the implementation of environmental policies at the local level in China. *Policy Research Working Paper, 7016*, 1-59.

Compilation of References

Kovel, J. (2000). The Struggle for Use Value. *Capitalism, Nature, Socialism, 11*(2), 3–23. doi:10.1080/10455750009358910

Kristof, N. (1997, January 9). For Third World water is still a deadly drink. *New York Times*, p. 2.

Kunmin, Z., Zongguo, W., & Liying, P. (2007). Environmental policies in China; Evolvement, features, and evaluation. China Population, Resources and Environment, 17(2), 1-7.

Kunnas, J. (2009). The Theory of Justice in A Warming Climate. *Earth Environ. Science,* (6), 11. doi:10.1088/1755-1307/6/1/112029

Label, L., Xu, J., Bastakoti, R. C., & Lamba, A. (2010). Pursuits of adaptiveness in the shared rivers of Monsoon Asia. *International Environmental Agreement: Politics, Law and Economics, 10*, 355–375. doi:10.100710784-010-9141-7

Lane, C. N. (2003). *Acid rain: overview and abstracts*. New York: Nova Science Publishers.

Laroche, M., Bergeron, J., & Barbaro-Forleo, G. (2001). Targeting consumers who are willing to pay more for environmentally friendly products. *Journal of Consumer Marketing, 18*(6), 503–520. doi:10.1108/EUM0000000006155

Lee, K. (2008). Opportunities for green marketing: Young consumers. *Marketing Intelligence & Planning, 26*(6), 573–586. doi:10.1108/02634500810902839

Leuenberger, D. Z., & Wakin, M. (2007). Sustainable Development in Public Administration Planning: An Exploration of Social Justice, Equity, and Citizen Inclusion. *Administrative Theory & Praxis Journal, 29*(3), 394–411.

Lewis, J. (2011). *Energy and Climate Goals of China's 12th Five-Year Plan*. Retrieved from http://www.c2es.org/international/key-country-policies/china/energy-climate-goals-twelfthfive-Year-plan

Li, L. (2013). An Exploration of the Maturation of Sino-Indian Relations and Its Causes [ZhongYinguanxi zouxiang chengshu ji qi yuanyin tanxi]. *Contemporary International Relations, 3*, 49–55.

Lipschutz, R. D. (1997). *Damming Troubled Waters: Conflict over the Danube: 1950-2000*. Paper presented at Environment and Security Conference, Institute of War and Peace Studies, New York, NY.

Liu, J., & Raven, P. H. (2010). China's environmental challenges and implications for the world. *Critical Reviews in Environmental Science and Technology*, 1–29.

Lockwood, B., & Redoano, M. (2005). *The CSGR Globalization Index: An Introductory Guide Centre for the Study of Globalization and Regionalization working paper155/04*. Retrieved from www2.warwick,ac.uk.fac/soc.csgr/index/citation

Longley, P. A., & Singleton, A. D. (2009). Classification through consultation: Public views of the geography of the E-Society. *International Journal of Geographical Information Science*, *23*(6), 737–763. doi:10.1080/13658810701704652

Lothar, B. (1991). Peace Through Parks: The Environment on the Peace Research Agenda. *Journal of Peace Research*, *28*(40), 407–423.

Magoulas, G., Lepouras, G., & Vassilakis, C. (2007). Virtual reality in the E-Society. *Virtual Reality Journal*, *11*(2), 71 – 73.

Mahapatra, S. K., & Ratha, K. C. (2016). Brahmaputra River: A bone of contention between India and China. *Water Utility Journal, 13*, 91-99. Retrieved from: http://www.ewra.net/wuj/pdf/WUJ_2016_13_08.pdf

Majláth, M. (2010). *Can Individuals do anything for the Environment? - The Role of Perceived Consumer Effectiveness.* Paper presented at the FIKUSZ '10 Symposium for Young Researchers, Budapest, Hungary.

Malthus, T. (1998). *An Essay on the Principle of Population*. London: Electronic Scholarly Publishing Project. Available at: http://129.237.201.53/books/malthus/population/malthus.pdf

Man, J. Y. (2013). *China's environmental policy and urban development*. Lincoln Institute of Land Policy.

Manaktola, K., & Jauhari, V. (2007). Exploring consumer attitude and behaviour towards green practices in the lodging industry in India. *International Journal of Contemporary Hospitality Management*, *19*(5), 364–377. doi:10.1108/09596110710757534

Margulis, S. (2004). *Causes of deforestation of the Brazilian Amazon* (Vol. 22). World Bank Publications.

Marks, P. (2005, June 8). *News Scientist*. Retrieved from www.newscientist.com: https://www.newscientist.com/article/dn7488-global-warming-is-a-clear-and-increasing-threat/

Martens, P., Akin, S., Maud, H., & Mohsin, R. (2010). Is globalization healthy: A statistical indicator analysis of the impacts of globalization on health. *Globalization and Health*, *6*(16). doi:10.1186/1744-8603-6-16 PMID:20849605

Martens, P., & Raza, M. (2009). Globalization in the 21st century: Measuring regional changes in multiple domains. *Integrated Assessment*, *9*(1), 1–18.

Martens, P., & Raza, M. (2010). Is globalization sustainable? *Sustainability*, *2*(1), 280–293. doi:10.3390u2010280

Martin, L. (1992). Interests, Power and Multilateralism. *International Organization*, *46*(4), 765–792. doi:10.1017/S0020818300033245

Mather, A. S., & Chapman, K. (1995). Environmental Resources. London: Longman.

Compilation of References

Matsuno, H. (2009). *China's environmental policy: Its effectiveness and suggested approaches for Japanese companies.* Nomura Research Institute.

McEachern, M. G., & Carrigan, M. (2012). Revisiting contemporary issues in green/ethical marketing: An introduction to the special issue. *Journal of Marketing Management, 28*(3-4), 189–194. doi:10.1080/0267257X.2012.666877

McGrew, A. (1992). *A Global Society?* London: Polity Press.

McKibbin, W. J., & Wilcoxen, P. J. (2002). The Role of Economics in Climate Change Policy. *The Journal of Economic Perspectives, 16*(2), 107–129. doi:10.1257/0895330027283

Meadows, D. (1972). *The Limits to Growth: A report for the Club of Rome's project on the predicament of mankind, Part 1 Club of Rome.* Potomac Associates.

Meadows, D. (2005). *Limits to Growth: The 30 year Update.* London: Earthscan.

Methodhome. (2017). *Green Glossary.* Retrieved July 18, 2017, from https://www.methodhome.com.au/beyond-bottle/green-glossary/

Miller, M. (1995). *The Third World in global environmental politics.* Boulder, CO: Lynne Rienner Publishers.

Mirza, M. N. (2016). *Indus water disputes and India-Pakistan relations* (Doctoral dissertation). Retrieved from: https://archiv.ub.uni-heidelberg.de/volltextserver/20915/1/Mirza%20PhD%20Dissertation%20for%20heiDOK.pdf

Mitchell, R. B. (2010). *International politics and the environment.* Los Angeles, CA: Sage Publications.

Mody, A. (2004). *What Is an Emerging Market?* IMF Working Paper, Research Department, WP/04/177, International Monetary Fund.

Moisander, J. (2007). Motivational complexity of green consumerism. *International Journal of Consumer Studies, 31*(4), 404–409. doi:10.1111/j.1470-6431.2007.00586.x

Momaya, K. (2011). Cooperation for competitiveness of emerging countries: Learning from a case of nanotechnology. *Competitiveness Review, 21*(2), 152–170. doi:10.1108/10595421111117443

Morelli, J. (2011). Environmental Sustainability: A Definition for Environmental Professionals. *Journal of Environmental Sustainability, 1*(1), 1–10. doi:10.14448/jes.01.0002

Muijs, D. (2010). *Doing quantitative research in education with SPSS.* Thousand Oaks, CA: Sage Publications.

Mustafa, K. (2014, January 21). India plans new dam on Chenab violating Indus water treaty. *The News.* Retrieved from: http://www.thenews.com.pk/Todays-News-2-227667-India-plans-new-dam-on-Chenab-violating-Indus-water-treaty

Mu, Z., Bu, S., & Xue, B. (2014). Environmental legislation in China; Achievements, challenges and trends. *Sustainability*, 1–13.

Najam, A. (2005). Developing countries and global environmental governance: From contestation to participation to engagement. *International Environmental Agreement: Politics, Law and Economics, 5*(3), 303–321. doi:10.100710784-005-3807-6

Namastesolar. (2017). *Operations & Maintanance*. Retrieved from http:// www. namastesolar.com/

Narain, S. (2010). *The Cancun End Game, Bad For Us Bad For Climate. IIED Outreach, A multi stake holder magazine on environment and sustainable development*. Retrieved from http://www.field.org.uk/files/outreach_cs_cancun_outcomes.pdf

Naz, F. (2014). Water: a cause of power politics in South Asia. *Water and Society II. WIT Transactions on Ecology and The Environment, 178*. Retrieved from: https://www.witpress.com/Secure/elibrary/papers/WS13/WS13009FU1.pdf

Nazaruk, P. A. (2016). Why Should We Take Care Of Nature? *HuffPost*. Retrieved from: https://www.huffingtonpost.com/pawel-alva-nazaruk/why-should-we-take-care-o_b_12170852.html

Newbelgium. (2017). *New Belgium Brewing Company*. Retrieved July 18, 2017, from http:// www. newbelgium.com/

Newman, P. (2016). Sustainable Urbanization: Four stages of Infrastructure Planning and Progress. *The Journal of Sustainable Urbanization Planning and Progress, 1*(1), 3–10.

Newman, P. W. (2015). Transport infrastructure and sustainability: A new planning and assessment framework. *Smart and Sustainable Built Environment, 4*(2), 140–153.

Nielsen. (2011). *The 'Green' gap between environmental concerns and the cash register*. Retrieved 13 May 2017 from http://www.nielsen.com/us/en/insights/news/2011/the-green-gap-between-environmental-concerns-and-the-cash-register.html

Niklas, P. (2015). The Evidence on Globalization. *World Economy, 38*(3), 509–552. doi:10.1111/twec.12174

Nordhaus, W. D. (2001). *After Kyoto: Alternative Mechanisms to Control Global Warming*. Academic Press.

Nqa. (2017). *EMAS: EU Eco-Management and Audit Scheme*. Retrieved July 19, 2017 from https: //www.nqa. com/en-gb/ certification/ standards/emas

O'Neill, J. (2001). *Building Better Global Economic BRICs*. Global Economics, Paper No: 66, Goldman Sachs. Retrieved from https://www.gs.com

Obama, B. (2009a). *Nobel Lecture*. Retrieved from http://nobelprize.org/nobel_prizes/peace/laureates/2009/obamalecture_ en.html

Compilation of References

Obama, B. (2009b). *Remarks by the President at United Nations Secretary General Ban Ki-Moons Climate Change Summit*. Retrieved from http://www.un.org/wcm/webdav/site/climatechange/shared/Docume

Obergassel, W. (2016). Phoenix from the Ashes — An Analysis of the Paris Agreement to the United Nations Framework Convention on Climate Change. Academic Press.

Oberthür, S., Viña, A., & Morgan, J. (2015). *Getting Specific On The 2015 Climate Change Agreement: Suggestions For The Legal Text With An Explanatory Memorandum*. Working Paper.

OECD. (2006). *Environmental performance review of China*. OECD.

OECD. (2008). *Handbook on constructing composite indicators: Methodology and user guide*. Paris: OECD Publishing.

OECD. (2013). What is the impact of globalization on the environment? In *Economic Globalization: Origins and consequences*. Paris: OECD Publishing; doi:10.1787/9789264111905-8-

OECD. (n.d.). *Country Statistical Profile: China*. Retrieved from http://www.oecd-ilibrary.org/economics/country-statistical-profile-china_csp-chn-table-en

Ogbodo, S. G. (2010). The Paradox of the Concept of Sustainable Development under Nigeria's Environmental Law. *Journal of Sustainable Development*, *3*(3). doi:10.5539/jsd.v3n3p201

Ohmae, K. (1995). *The End of the Nation State*. New York: Free Press.

Ometto, J., Aguiar, A., & Martinelli, L. (2014). Amazon deforestation in Brazil: Effects, drivers and challenges. *Carbon Management*, *2*(5), 575–585. doi:10.4155/cmt.11.48

Otegbulu, A., & Adewunmi, Y. (2009). Evaluating the sustainability of urban housing development in Nigeria through innovative infrastructure management. *International Journal of Housing Markets and Analysis*, *2*(4), 334–346. doi:10.1108/17538270910992782

Ozaki, R. (2011). Adopting sustainable innovation: What makes consumers sign up to green electricity? *Business Strategy and the Environment*, *20*(1), 1–17. doi:10.1002/bse.650

Pagiaslis, A., & Krontalis, A. K. (2014). Green Consumption Behavior Antecedents: Environmental Concern, Knowledge, and Beliefs. *Psychology and Marketing*, *31*(5), 335–348. doi:10.1002/mar.20698

Pak, J. H. (2016). Challenges in Asia: China, India, and War over Water. *Parameters*, *46*(2). Retrieved from http://ssi.armywarcollege.edu/pubs/parameters/issues/Summer_2016/8_Pak.pdf

Pantelic, D., Sakal, M., & Zehetner, A. (2016). Marketing and sustainability from the perspective of future decision makers. *South African Journal of Business Management*, *47*(1), 37–47.

Parsai, G. (2010, June 1). India, Pakistan resolve Baglihar dam issue. *The Hindu*. Retrieved from: http://www.thehindu.com/news/India-Pakistan-resolve-Baglihar-dam-issue/article16240199.ece

Pataki, G. E., & Crotty, E. M. (2017). *Understanding and Implementing an Environmental Management System: A step by step guide.* New York State Department of Environmental Conversation Pollution Prevention Unit. Retrieved July 18, 2017, from http://www.dec.ny.gov/docs/permits_ej_operations_pdf/p2emsstep1.pdf

Paul, J., Modi, A., & Patel, J. (2016). Predicting green product consumption using theory of planned behavior and reasoned action. *Journal of Retailing and Consumer Services, 29,* 123–134. doi:10.1016/j.jretconser.2015.11.006

Perron-Welch, F. (2011). *The Future of Global Forests after the Cancun Climate Change Conference IDLO Sustainable development law on Climate Change.* Legal working paper series.

Petheram, L. (2002). *Our Planet in Peril: Acid rain.* Bridgestone Books.

Pickering, K. T., & Owen, L. A. (2006). *An introduction to global environmental issues* (2nd ed.). London: Routledge.

Popescu, F., & Lonel, I. (2010). Anthropogenic air pollution sources. In *Air quality.* InTech. doi:10.5772/9751

Pour-Ghaz, M. (2013). Sustainable Infrastructure Materials: Challenges and Opportunities. *International Journal of Applied Ceramic Technology, 10*(4), 584–592. doi:10.1111/ijac.12083

Presentation at the 20th Anniversary Meeting of the International Energy Workshop, Romano, GC. (2010). *The EU-China Partnership on Climate Change: Bilateralism Begetting Multilateralism in Promoting a Climate Change Regime?* Mercury, paper No 8. Pp 1-28.

Purdey, S. J. (2010). *Economic growth, the environment and international relations: the growth paradigm.* London: Routledge.

Qin, B., Liu, Z., & Havens, K. (Eds.). (2007). *Eutrophication of shallow lakes with special reference to Lake Taihu, China* (Vol. 194). London: Springer. doi:10.1007/978-1-4020-6158-5

Qin, X. W. (2015). *China's path to a green economy: Decoding China's green economy concepts and policies.* International Institute for Environment and Development.

Qi, Y. (2007). *China's Challenges in environmental regulation.* School of Public Policy and Management: Tsinghua University.

Raiyn, J. (2011). Developing E-Society Cognitive Platform Based on the Social Agent E-learning Goal Oriented. *Advances in Internet of Things, 1*(1), 1–4. doi:10.4236/ait.2011.11001

Rashid, N. R. (2009). Awareness of eco-label in Malaysia's green marketing initiative. *International Journal of Business and Management, 4*(8), 132–150. doi:10.5539/ijbm.v4n8p132

Rediff. (2005). *Baglihar verdict to be binding on Pak, India: Swiss Expert.* Retrieved from: http://www.rediff.com/news/2005/oct/04baglihar.htm

REN21. (2014). *Renewables 2014 Global Status Report.* Retrieved from http://www.ren21.net/Portals/0/documents/Resources/GSR/2014/GSR2014_full%20report_low%20res.pdf

Compilation of References

Rennen, W., & Martens, P. (2003). The globalization timeline. *Integrated Assessment*, *4*(3), 137–144. doi:10.1076/iaij.4.3.137.23768

Rex, E., & Baumann, H. (2007). Beyond ecolabels: What green marketing can learn from conventional marketing. *Journal of Cleaner Production*, *15*(6), 567–576. doi:10.1016/j.jclepro.2006.05.013

Rio Report, U. N. (2012). *The Future We Want*. Oxford Press.

Roberts, D. (2008). Thinking globally, acting locally—institutionalizing climate change at the local government level in Durban, South Africa. *Environment and Urbanization*, *20*(2), 521–537. doi:10.1177/0956247808096126

Roberts, J. A. (1996). Green consumers in the 1990s: Profile and implications for advertising. *Journal of Business Research*, *36*(3), 217–231. doi:10.1016/0148-2963(95)00150-6

Roberts, J., & Thanos, N. (Eds.). (2003). *Trouble in Paradise: Globalization and Environmental Crises in Latin America*. New York: Routledge.

Robertson, R. (1992). *Globalization: Social Theory and Global Culture*. Sage.

Roe, B., Teisl, M. F., Rong, H., & Levy, A. S. (2001). Characteristic of consumer-preferred labelling policies: Experimental evidence from price and environmental disclosure for deregulated electricity services. *The Journal of Consumer Affairs*, *35*(1), 1–26. doi:10.1111/j.1745-6606.2001.tb00100.x

Rogers, K. (1997). *Ecological Security and Multinational Corporations*. Kluwer Academic Publishers.

Rosa, D. J. (2009). Sustainability and Infrastructure Resource Allocation. *Journal of Business & Economics Research*, *7*(9).

Rosenau, J. (1996). *Complexities and Contradictions of Globalization. World Politics*. Cambridge, UK: Cambridge University Press.

Rousseau, G., & Venter, D. (2001). A multi-cultural investigation into consumer environmental concern. *SA Journal of Industrial Psychology*, *27*(1), 1–7. doi:10.4102ajip.v27i1.768

Rozelle, S., Huang, J., & Zhang, L. (1997). Poverty, population and environmental degradation in China. *Food Policy*, *22*(3), 229–251.

Ruggie, J. G. (1992). Multilateralism: The Anatomy of an Institution. *International Organization*, *3*(46), 561–598. doi:10.1017/S0020818300027831

Ryan, B. (2006). Green Consumers A Growing Market for Many Local Businesses. *UWExtension*, *1*(123), 1-2.

Samaranayake, N., Limaye, S., & Wuthnow, J. (2016). *Water Resource Competition in the Brahmaputra River Basin: China, India, and Bangladesh*. Washington, DC: CNA Analysis and Solution. Retrieved from https://www.cna.org/cna_files/pdf/cna-brahmaputra-study-2016.pdf

Scheren, P., Ibe, A., Janssen, F., & Lemmens, A. (2002). Environmental pollution in the Gulf of Guinea – a regional approach. *Marine Pollution Bulletin*, *44*(7), 633–641. doi:10.1016/S0025-326X(01)00305-8 PMID:12222886

Schmid, B., Stanoevska, K., & Tschammer, V. (2013). *Towards the E-Society: E-Commerce, E-Business, and E-Government*. Springer Publishing Company, Incorporated.

Schor, J. (2001). *Why Do We Consume So Much?* Saint John's University Clemens Lecture Series.

Schor, J. (2005). Prices and quantities: Unsustainable consumption and the global economy. *Ecological Economics*, *55*(3), 309–320. doi:10.1016/j.ecolecon.2005.07.030

Scott, L., & Vigar-Ellis, D. (2014). Consumer understanding, perceptions and behaviours with regard to environmentally friendly packaging in a developing nation. *International Journal of Consumer Studies*, *38*(6), 642–649. doi:10.1111/ijcs.12136

Seventhgeneration. (2017). *Nurture Nature*. Retrieved July 18, 2017, from https:// www. seventh generation. com/nurture-nature

Shanaha, M. (2010). *Journalist from Climate Change Frontline Kept World's Eyes Focused on COP 16. IIED Outreach, A multi stake holder magazine on environment and sustainable development*. Retrieved from http://www.field.org.uk/files/outreach_cs_cancun_outcomes.pdf

Sharma, S. P. (2006). Indo-Pak pact on waters hurdle in power projects. *Tribune News Service*. Retrieved from: http://www.tribuneindia.com/2005/20051008/j&k.htm#1

Sharma, Y. (2011). Changing consumer behaviour with respect to green marketing–a case study of consumer durables and retailing. *International Journal of Multidisciplinary Research*, *1*(4), 152–162.

Sheldon, C., & Yoxon, M. (2006). *Environmental management systems: a step-by-step guide to implementation and maintenance* (3rd ed.). Routledge.

Shen, L., Wu, Y., & Zhang, X. (2011). Key Assessment Indicators for the Sustainability of Infrastructure Projects. *Journal of Construction Engineering and Management, Publisher American Society of Civil Engineers*, *137*(6).

Shiva, V. (1988). *Staying alive: Women, ecology, and development*. London: Zed Books.

Shortle, J. S., & Abler, D. G. (Eds.). (2001). *Environmental policies for agricultural pollution control*. New York: CABI Publishing. doi:10.1079/9780851993997.0000

Shove, E. (2003). Converging Conventions of Comfort, Cleanliness and Convenience. *Journal of Consumer Policy*, *26*(4), 395–418. doi:10.1023/A:1026362829781

Siew, R. Y. J., Balatbat, M. C. A., & Carmichael, D. G. (2016). A proposed framework for assessing the sustainability of infrastructure. *International Journal of Construction Management*.

Simon, Y. (1997). Encyclopédie des marchés financiers (Vols. 1-2). Editeur Economica.

Compilation of References

Sindico. (2010). The Copenhagen Accord and the Future of the International Climate Change Regime. *Revista Catalana de Dret Ambiental, 1*(1), 1 – 24.

Singer, P. (2002). *One World: The Ethics of Globalization.* New Haven, CT: Yale University Press.

Sinnappan, P., & Rahman, A. A. (2011). Antecedents of green purchasing behavior among Malaysian consumers. *International Business Management, 5*(3), 129–139. doi:10.3923/ibm.2011.129.139

Smith, A. (2009). *The Nobel Peace Prize 2009: Time for Hope.* Retrieved from http://nobelprize.org/nobel_prizes/peace/laureates/2009/speedread.html

Smith, E. E., & Perks, S. (2010). A perceptual study of the impact of green practice implementation on the business functions. *Southern African Business Review, 14*(3), 1–29.

Smith, G. L., Sundaram, S. K., & Spearing, D. R. (2002). *Environmental Issues and Waste Management Technologies in the Ceramic and Nuclear Industries VII, Ceramic Transactions* (Vol. 132). The American Ceramic Society.

Smith, J. (2017). How are GMOs and Roundup linked to cancer? *International Conference on The Truth About Cancer Orlando United States.*

Solberg. (2017). *Serious about Sustainability.* Retrieved July 18, 2017, from http://www.solbergmfg.com/

Somlyódy, L., & van Straten, G. (Eds.). (2012). *Modeling and managing shallow lake eutrophication: with application to Lake Balaton.* London: Springer.

Soyez, K. (2012). How national cultural values affect pro-environmental consumer behavior. *International Marketing Review, 29*(6), 623–646. doi:10.1108/02651331211277973

Spaargaren, G. (2003). Sustainable Consumption: A Theoretical and Environmental Policy Perspective. *Society & Natural Resources, 16*(8), 687–701. doi:10.1080/08941920309192

Spaargaren, G., & Mol, A. (1992). Sociology, Environment and Modernity: Ecological Modernization as a Theory of Social Change. *Society & Natural Resources*, • • •, 5.

Speake, S., & Gismondi, M. (2005). Water: A Human Right. In *Consuming Sustainability Critical Social Analysis of Ecological Change.* Fernwood Publishing.

Springer, U. (2002). The market for trade able GHG permits under the Kyoto Protocol: A survey of model studies. *Energy Economics, 25*(5), 527–551. doi:10.1016/S0140-9883(02)00103-2

Statistics, S. A. (2011). Msunduzi Municipanilty. *Statistics SA.* Retrieved 13 May 2017 from http://www.statssa.gov.za/?page_id=993&id=the-msunduzi-municipality

Stavins, R. N. (2010). *A Look Back At Cop-16. What happened (and why) in Cancun. IIED Outreach, A multi stake holder magazine on environment and sustainable development.* Retrieved on September, 28, 2011 from the website http://www.field.org.uk/files/outreach_cs_cancun_outcomes.pdf

Stern, P. (1997). *Environmentally Significant Consumption. Research Directions.* National Academy Press.

Stern, P. C. (2000). New environmental theories: Toward a coherent theory of environmentally significant behavior. *The Journal of Social Issues, 56*(3), 407–424. doi:10.1111/0022-4537.00175

Stiglitz, J. (2003). *Globalization and Its Discontents.* London: Cambridge University Press.

Stiglitz, J. (2016). *Globalization and its New Discontents.* Project Syndicate.

Straughan, R. D., & Roberts, J. A. (1999). Environmental segmentation alternatives: A look at green consumer behavior in the new millennium. *Journal of Consumer Marketing, 16*(6), 558–575. doi:10.1108/07363769910297506

Strode, A., Slack, C., & Essack, Z. (2010). Child consent in South African law: Implications for researchers, service providers and policy-makers. *SAMJ: South African Medical Journal, 100*(4), 247–249. doi:10.7196/SAMJ.3609 PMID:20459973

Sulphey, M. M., & Safeer, M. M. (2015). *Introduction to Environment Management* (3rd ed.). Delhi: PHI Learning Pvt. Ltd.

Sungevity. (2017). *Solar Basics.* Retrieved from http: //www. sungevity.com/home/ solar-panel-savings?_ga= 2.224662425.285111617.1500500164-1897408410. 1500394429

Sunpower. (2017). *Positive Energy Solar.* Retrieved from https:// www. positiveenergysolar. com/

Sustainable Development Knowledge Platform, Division For Sustainable Development, UN-DESA. (n.d.). The United Nations, Department of Economic and Social Affairs. Retrieved from https://sustainabledevelopment.un.org/

Swilling, M. (2006). Sustainability and infrastructure planning in South Africa: A Cape Town case study. *Environment and Urbanization, 18*(1), 23–50. doi:10.1177/0956247806063939

Synodinos, C., & Bevan-Dye, A. (2014). Determining African Generation Y Students' Likelihood of Engaging in Pro-environmental Purchasing Behaviour. *Mediterranean Journal of Social Sciences, 5*(21), 101–110.

Takács-Sánta, A. (2007). Barriers to environmental concern. *Human Ecology Review, 14*(1), 26.

Tan, S. (2014). *Challenges and Issues of Sustainable Development Policies.* Lee Kuan Yew School of Public Policy Research Paper No. 14-25. Available at SSRN: http://ssrn.com/abstract=2484930

Tang, X., McLellan, B. C., Snowden, S., & Hook, M. (2015). Dilemmas for China; Energy, Economy and Environment. *Sustainability,* 1-13.

Terhune, C. (2004, June 18). Coke CEO Says obesity is a challenge. *Wall Street Journal.*

The Berne Declaration. (2013). *Agropolicy. A handful of corporations control world food production.* Available at http://www.econexus.info/sites/econexus/files/Agropoly_Econexus_BerneDeclaration

Compilation of References

The Express Tribune. (2016, October 1). *China blocks Brahmaputra River as India threatens to scrap Indus Water Treaty*. Retrieved from: https://tribune.com.pk/story/1191953/china-blocks-brahmaputra-river-india-threatens-scrap-indus-water-treaty/

The Express Tribune. (2017, February 26). *India fails to substantiate claim of Pakistan's involvement in Uri attack*. Retrieved from: https://tribune.com.pk/story/1339596/india-fails-substantiate- claim-pakistans-involvement-uri-attack/

The Nation. (2016, October 17). *India's suicidal water pursuits*. Retrieved from: http://nation.com.pk/17-Oct-2016/india-s-suicidal-water-pursuits

The News. (2016, October 2). *China stops water of Brahmaputra River tributary*. Retrieved from: https://www.thenews.com.pk/print/154331-China-stops-water-of-Brahmaputra-River-tributary

The News. (2017, June 10). *India completes Kishanganga hydropower project without resolving differences*. Retrieved from: https://www.thenews.com.pk/print/155112-India-completes-Kishanganga- hydropower-project-without-resolving-differences

The Times of India. (2016, April 2). *Pakistan claims India 'failed' to provide evidence on Pathankot attack: Report*. Retrieved from: https://timesofindia.indiatimes.com/india/Pakistan-claims-

Thøgersen, J., & Ölander, F. (2003). Spillover of environment-friendly consumer behaviour. *Journal of Environmental Psychology*, *23*(3), 225–236. doi:10.1016/S0272-4944(03)00018-5

Thurow, L. (1992). *The Coming Economic Battle Among Japan America, and Europe*. New York: William Marrow and Company Inc.

Tilikidou, I. (2007). The effects of knowledge and attitudes upon Greeks' pro-environmental purchasing behaviour. *Corporate Social Responsibility and Environmental Management*, *14*(3), 121–134. doi:10.1002/csr.123

Tinsley, S. (2002). EMS models for business strategy development. *Business Strategy and the Environment*, *11*(6), 376–390. doi:10.1002/bse.340

Tiwary, A., & Colls, J. (2010). *Air pollution: Measurement, modeling, and mitigation* (3rd ed.). London: Taylor & Francis.

Tobler, C., Visschers, V. H., & Siegrist, M. (2012). Addressing climate change: Determinants of consumers' willingness to act and to support policy measures. *Journal of Environmental Psychology*, *32*(3), 197–207. doi:10.1016/j.jenvp.2012.02.001

Today, P. (2017, February 3). *UNDP report on water security: A wake up call*. Retrieved from: https://www.pakistantoday.com.pk/2017/02/03/undp-report-on-water-security/

Tomlinson, J. (1999). *Globalization and Culture*. University of Chicago Press.

Turner, R. K. (1988). *Sustainable environmental management*. London: Belhaven, UNFCCC 1992. U.N. Doc. A/AC.237/18, reprinted in 31 I.L.M. 849.

Ugwu, O. O. (2013). Nanotechnology for sustainable infrastructure in 21st century civil engineering. *Journal of Innovative Engineering*, *1*(1).

UN. (2002a). *Johannesburg declaration on sustainable development*. United Nations.

UNDP - Human Development Report. (1998). *Brazil -Globalization and Changes in Consumer Patterns*. Oxford Press.

UNEP (United Nations Environment Programme). (2011). Environmental Assessment of Ogoniland. UNEP Report.

UNEP. (2017). *UNEP / FIDIC / ICLEI Urban Environmental Management: Environmental Management Training Resources Kit*. Retrieved July 18, 2017, from http: //www. unep.or.jp / ietc/Announcements /EMSkit_launch.asp

UNESCO. (n.d.). *UNESCO*. Retrieved from www.unesco.org

UNFCCC. (2009). *Copenhagen Accord*. Retrieved from http://www.denmark.dk/NR/rdonlyres/C41B62AB-4688-4ACE-BB7BF6D2C8AAEC20/ 0/copenhagen_accord.pdf

United Nations (UN). (1997). Commission on Sustainable Development. Comprehensive assessment of the freshwater resources of the world. New York: UN.

United Nations Development Programme (UNDP). (n.d.a). *Sustainable Development Goals: Background on the Goals*. Retrieved from http://www.undp.org/content/undp/en/home/sustainable-development-goals/background.html

United Nations Development Programme (UNDP). (n.d.b). *Sustainable Development Goals*. Retrieved from http://www.undp.org/content/undp/en/home/sustainable-development-goals.html

United Nations Development Programme. (1994). *Human Development Report: Redefining Security*. Oxford, UK: Human Dimensions.

United Nations Environment Programme, Division of Technology, Industry and Economics. (2011). *Paving the Way for Sustainable Consumption and Production: The Marrakech Process Progress report*. Available at: http://www.unep.fr/scp/marrakech/pdf/Marrakech%20Process%20Progress%20Report%20FINAL.pdf

United Nations, Report of the World Summit on Sustainable Development. (2002). Available at: http://www.johannesburgsummit.org/html/documents/summit_docs/131302_wssd_report_reissued.pdf

United Nations. (1987). *Our Common Future. Report of the World Commission on Environment and Development*. Oxford Press.

United Nations. (1992). *United Nations Sustainable Development, Earth Summit: Agenda 21, Chapter 4.3*. Available at: http://www.un.org/esa/sustdev/documents/agenda21/english/Agenda21.pdf

Compilation of References

United Nations. (2012). *The Future We Want*. Available at: http://www.uncsd2012.org/content/documents/727The%20Future%20We%20Want%2019%20June%201230pm.pdf

United Nations. (2015). *World Population Prospects: The 2015 Revision*. Retrieved 13 May 2017 from: https://populationpyramid.net/world/2020/

United Nations. (2016). *Environment Annual Report 2016*. Retrieved 13 May 2017 from: http://web.unep.org/annualreport/2016/index.php

United Nations. (n.d.). *United Nations*. Retrieved from http://www.un.org/

US Energy Information Administration. (2013). Total Carbon Dioxide Emissions from the Consumption of Energy (Million Metric Tons). *International Energy Outlook 2013*.

Vajpeyi, D. K. (2001). *Deforestation, environment, and sustainable development: a comparative analysis*. Greenwood Publishing Group.

van Riper, C. J., & Kyle, G. T. (2014). Understanding the internal processes of behavioral engagement in a national park: A latent variable path analysis of the value-belief-norm theory. *Journal of Environmental Psychology*, *38*, 288–297. doi:10.1016/j.jenvp.2014.03.002

van Santen, L. (2013). *Timeline: a brief history of sustainable consumption*. World Economic Forum. Retrieved from: https://www.weforum.org/agenda/2013/11/timeline-a-brief-history-of-sustainable-consumption/

Varinsky, D. (2017). 5 claims Trump used to justify pulling the US out of the Paris Agreement—and the reality, Tech. *Business Insider*. Retrieved from http://www.pulse.com.gh/bi/tech/tech-5-claims-trump-used-to-justify-pulling-the-us-out-of-the-paris-agreement-and-the-reality-id6773413.html

Veseth, M. (2004). *What is International Political Economy?* UNESCO.

Vezzoli, C. A., & Manzini, E. (2008). *Design for Environmental Sustainability*. Milan: Springer Science & Business Media.

Vicente-Molina, M. A., Fernández-Sáinz, A., & Izagirre-Olaizola, J. (2013). Environmental knowledge and other variables affecting pro-environmental behaviour: Comparison of university students from emerging and advanced countries. *Journal of Cleaner Production*, *61*, 130–138. doi:10.1016/j.jclepro.2013.05.015

Vikan, A., Camino, C., Biaggio, A., & Nordvik, H. (2007). Endorsement of the New Ecological Paradigm A Comparison of Two Brazilian Samples and One Norwegian Sample. *Environment and Behavior*, *39*(2), 217–228. doi:10.1177/0013916506286946

Vitousek, P., Ehrlich, P., Ehrlich, A., & Matson, P. (1986). Human appropriation of the products of photosynthesis. *Bioscience*, *36*(6), 368–373. doi:10.2307/1310258

Wackernagel, M., Onisto, L., Bello, P., Callejas Linares, A., Susana López Falfán, I., Méndez García, J., ... Guadalupe Suárez Guerrero, M. (1999). National natural capital accounting with the ecological footprint concept. *Ecological Economics*, *29*(3), 375–390. doi:10.1016/S0921-8009(98)90063-5

WasteZero. (2017). *The Trash Problem*. July 18, 2017, from http:// wastezero. com/

Waterman, S. (1984). Partition—A Problem in Political Geography. In P. Taylor & J. House (Eds.), *Political Geography*. London: Croom Helm.

WCED (World Commission on Environment and Development). (1987). Our Common Future. Oxford University Press.

Wehmier, S., & Ashby, M. (Eds.). (2000). *Oxford Advanced Learner's Dictionary*. Oxford, UK: Oxford University Press.

Wehrli, A. (2014). Why Mountains Matter for Sustainable Development. *Mountain Research and Development*, *34*(4), 405–409. doi:10.1659/MRD-JOURNAL-D-14-00096.1

Weiß, P., & Bentlage, J. (2006). Environmental Management Systems and Certification. Baltic University Press.

Werksman, J. (1995). Greening Bretton Woods. In The Earthscan Reader in Sustainable Development. London: Earthscan.

Wessels, M. (2014) Stimulating sustainable infrastructure development through public–private partnerships. *Proceedings of the Institution of Civil Engineers - Management, Procurement and Law*, *167*(5), 232 – 241.

WestPaw. (2017). *Eco-friendly materials*. Retrieved from https: //www. westpaw. com/

White, L. (1967). The historical roots of our ecological crisis. *Science*, *155*(3767), 1203–1207. doi:10.1126cience.155.3767.1203 PMID:17847526

WHO. (2001). *Environment and People's health in China*. United Nations Development Programme.

Williams, L. (2014). *China's climate change policies*. Lowy Institute.

Wirsing, R. (2008). The Kahsmir territorial dispute: The Indus runs through it. *The Brown Journal of World Affairs*, *15*, 225–240.

Wise, T. (2015). Two Roads Diverged in the Food Crisis: Global policy takes the one moreq traveled. *Canadian Food Studies*, *2*(2), 1-15.

World Bank. (2015). World Development Indicators. Washington, DC: World Bank.

World Bank. (n.d.). *The Indus Waters Treaty (IWT)*. Retrieved on July 23, 2017 from: http://siteresources.worldbank.org/INTSOUTHASIA/Resources/223497- 1105737253588/IndusWatersTreaty1960.pdf

Compilation of References

World Bank. (n.d.). *World Bank Open Data*. Retrieved from http://data.worldbank.org/

World Commission in Economic Development. (1987). *Report of the World Commission on Environment and Development: Our Common Future*. Available at: http://www.un-documents.net/our-common-future.pdf

World Economic Forum (WEF). (2013). *Sustainable Consumption: Stakeholder Perspectives*. Retrieved from: http://www3.weforum.org/docs/WEF_ENV_SustainableConsumption_Book_2013.pdf

World Health Organization (WHO). (n.d.). *Millennium Development Goals (MDGs)*. Retrieved from: http://www.who.int/topics/millennium_development_goals/about/en/

World Health Organization. (1997). Health and environment in sustainable development: Five years after the earth summit. Geneva: World Health Organization. (No. WHO/EHG/97.8)

Worldwatch. (n.d.). *Environmental Milestones: A Worldwatch Institute timeline tracing key moments in the sustainability movement from 1960s to 2004*. Retrieved from: https://www.worldwatch.org/brain/features/timeline/timeline.htm

Worzel, R. (1994). *Facing the Future; The Seven Forces Revolutionizing Our Lives*. Stoddart Publishing.

Wu, J., Deng, Y., Huang, J., Morck, R., & Yeung, B. (2013). *Incentives and outcomes; China's environmental policy*. National University of Singapore.

Wu, T. (2011, November 7). *The Diplomat*. Retrieved from www.thediplomat.com: http://thediplomat.com/2011/11/china-brics-and-the-environment/

Xinhua News Agency. (n.d.). *Xinhua Net*. Retrieved from http://www.xinhuanet.com/english/

Xu, H. (1998). Environmental policy and rural industrial development in China. *Research in Human Ecology*, *6*(2), 72–80.

Yadav, R., & Pathak, G. S. (2016). Intention to purchase organic food among young consumers: Evidences from a developing nation. *Appetite*, *96*, 122–128. doi:10.1016/j.appet.2015.09.017 PMID:26386300

Yakhou, M., & Dorweiler, V. P. (2004). Environmental accounting: An essential component of business strategy. *Business Strategy and the Environment*, *13*(2), 65–77. doi:10.1002/bse.395

Yasmine, R. (2010). COP 15 and Pacific Island States: A Collective Voice on Climate Change. *Pacific Journalism Review*, *16*(1), 193–203.

Young, A. (2015, March 22). *IBTImes*. Retrieved from www.ibtimes.com: http://www.ibtimes.com/global-warming-china-wealth-accumulation-will-amplify-climate-disasters-china-says-1855000

Young, W., Hwang, K., McDonald, S., & Oates, C. J. (2010). Sustainable consumption: Green consumer behaviour when purchasing products. *Sustainable Development*, *18*(1), 20–31.

Zawahri, N. (2009). Third party mediation of international river disputes: Lessons from the Indus river. *International Negotiation*, *14*(2), 281–310. doi:10.1163/157180609X432833

Zhang, H. (2015). *China-India Water Disputes: Two Major Misperceptions Revisited*. S. Rajaratnam School of International Studies, Nanyang Technological University, 015. Retrieved from: https://www.rsis.edu.sg/wp-content/uploads/2015/01/CO15015.pdf

Zhang, H. (2016). Sino-Indian water disputes: the coming water wars? *Wiley Interdisciplinary Reviews: Water WIREs Water*, *3*, 155–166. Retrieved from: http://onlinelibrary.wiley.com/doi/10.1002/wat2.1123/pdf

Zhang, Q., Crooks, R., & Jiang, Y. (2011). *Environmentally sustainable development in the People's Republic of China; Vision for the future and the rold of the Asian Development Bank*. Asian Development Bank.

Zheng, Y., & Walsham, G. (2008). Inequality of what? Social exclusion in the E-Society as capability deprivation. *Information Technology & People*, *21*(3), 222–243. doi:10.1108/09593840810896000

Zhifei, L. (2013). Water Security Issues in Sino-Indian Territorial Disputes [ZhongYin lingtu zhengduan zhong de shui ziyuan anquan wenti]. *South Asian Studies Quarterly*, *4*, 29–34.

Zhu, J., Yan, Y., & He, C. (2015). *China's environment; Big issues, accelerating effort, ample opportunities*. Equity Research.

Zhu, J., Yan, Y., He, C., & Wang, C. (2015). *China's environment; Big issues, accelerating effort, ample opportunities*. Goldman Sachs: Equity Research.

About the Contributors

Sofia Idris has done MS Gender Studies from University of the Punjab (Pakistan) and M Phil Political Science from GC University, Lahore (Pakistan). Her areas of interest include local politics, current affairs, political economy, international relations, conflict and conflict resolution, society & social systems, social & cultural issues, social care, women empowerment, children and gender issues.

* * *

Luke Amadi is a Senior Research Fellow at the Educational Support and Development Initiative for the Less Privileged(ESDIL). He specialized in Development Studies from the University of Port Harcourt, Nigeria and is published in the areas of Political Ecology, Sustainable Development, International Political Economy, Livelihoods, Globalization, Culture Studies and Identity. He has several publications in local and international scholarly journals. His recent publication is Intellectual capital and environmental sustainability measurement nexus: a review of the literature. Int. J. Learning and Intellectual Capital, Vol. X14 No. 2,92017) Inderscience, Geneva(Co-authored).

Favour Urenma Duruji-Moses holds a Masters Degree in International Relations and also manages Destewards Enterprises in Ota, Nigeria. She attended Covenant University where she obtained both bachelor and masters degrees. She has participated in some conferences within Nigeria and some of her works have been presented at local and international conferences. She has published articles in reputable outlets.

Moses Metumara Duuji is currently teaching at the Department of Political Science and International Relations, Covenant University, Ota Nigeria. He holds a PhD in Political Science as well as PGD in Journalism from the Nigerian Institute of Journalism Lagos. Dr. Duruji won a small grant of Harry Frank Guggenheim Foundation program for Young African Scholar in 2006; in 2008 the African Institute of South Africa gave him awards of Best Young Scholar and Best Innovative Researcher in 2008. His research interests are in the areas of Nationalism, Ethnicity, Democracy, Governance, Civil society, Federalism, and Globalization. His hobbies include reading, adventure and games.

Debbie Ellis is an Associate Professor in Marketing in the School of Management, IT and Governance, UKZN. She has a PhD from KTH Royal Institute of Technology, Sweden and a Masters in business Science from the University of Cape Town, South Africa. Debbie has 25 years of teaching, supervision and research experience. Her work has been published in Business Horizons, the International Journal of Wine Business Research, the Journal of Wine research, the Southern African Business Review, the South African Journal of Business Management, South African Journal of Economic and Management Sciences, and the International Journal of Consumer Studies among others. Her primary areas of research interest include eco-consumption, social marketing, young consumers, consumer knowledge, wine marketing, sports marketing and positioning.

Fauzia Ghani was a student of GC Univeristy, Lahore and started her career from this very institution as well. It has been 17 years since she started her work in this institution.

Prince Ikechukwu Igwe holds a PhD in Development Studies from the University of Port Harcourt Nigeria. He has several publications in scholarly journals.

Fatma İnce was born in Mersin, and spent their early life in the Mediterranean region, in the South of Turkey. They received a Ph. Degree in management and organization-business from the University of Erciyes, Turkey, in 2011. From 2007 to 2013, they worked in a vocational college as a department head and lecturer. After leaving the Cappadocia region, they came back to Mersin and joined the Department of Business Information Management, University of Mersin, as an Assistant Professor in 2014. Their current research interests include environment, entrepreneurship, leadership and organizational behaviors."

Njabulo Mkhize, MCom, completed his Master of Commerce in Marketing Management at the University of KwaZulu-Natal with a specific focus on green consumer behaviour and the green gap. He intends to continue with exploring this topic in his PhD studies. He is currently lecturing at the University of Johannesburg in Gauteng, South Africa where he lectures in various subjects in marketing management. He is also currently supervising students at post graduate level. He

About the Contributors

is passionate about educating the next generation of students in marketing and is an emerging researcher in the field and hopes to contribute immensely in the saving of the plant through providing empirical research in consumer's behaviour and organisation's role in a more sustainable environment.

Faith Osasumwen Olanrewaju is a Doctoral Student and Lecturer in the Department of Political Science and International Relations, Covenant University, Ota, Ogun State, Nigeria. She started teaching and research work in the university system in 2009 with research focus on Gender Issues, Development, Terrorism, National Security and Displacement. She has participated in some conferences within Nigeria and some of her works have been presented at local and international conferences. She has published articles in reputable local and international journals.

Komal Ashraf Qureshi completed her M Phil in International Relations at Kinnaird College for Women. She is currently working as a Graduate Teaching Fellow in Lahore School of Economics and works in the field of Research for both self-interest and better exploration of the events happening in the world. She is passionate about delivering effective knowledge to the academic and governmental institutions of Pakistan.

Index

A

Airpocalypse 137, 139, 141, 150, 153
anthropogenic greenhouse gases 32
Arid Agriculture 84
average temperature 5-6, 32

B

businesses 86-87, 89, 95, 102-103, 107, 112-113, 125-127, 138, 151

C

carbon dioxide 2, 5, 17, 32, 36, 94, 96, 98, 137, 139-146, 149, 156
carbon dioxide emissions 17, 32, 139
Chenab 59, 63-64, 66-69, 74, 77, 84
climate change 2, 4-5, 11, 21, 31-32, 34-36, 39-41, 44-51, 62-63, 71-72, 75, 78, 92-95, 105, 112-113, 117, 144-145, 177
committees 68, 86, 88, 93, 95, 107
conflict 51, 59, 61-62, 65, 68, 70-71, 74-75, 78-79, 84, 91-92, 145, 150

D

deforestation 11, 19, 23, 42-44, 86, 92, 94, 97-98, 107, 113, 139, 142, 145, 150, 152
Developing Economies 162-164, 168-169, 178, 183
digital 163-164, 173-174, 183

E

Eco-Labels 124, 135
Ecological Conscious Consumer Behavior (ECCB) 135
Ecological Footprint 20, 30
ecological justice 2, 5, 14-15, 20-22, 30
ecological modernization 8, 15, 18, 30
economic growth 21, 61, 70-71, 93, 138, 146, 150, 152, 165-166, 168, 172-173, 176, 178-179
emerging countries 41, 163-164, 168-169, 171, 173-174, 178-179, 183
emerging market 112, 114, 118, 126-127, 171
emission 11, 36-41, 43, 45-49, 86, 94-95, 104, 137, 139-144, 146-147, 149-150, 152, 156
environmental concern 113-116, 118, 121-123, 126, 135
environmental issues 17, 62, 69-70, 75, 78, 86-88, 91-94, 101-103, 107, 113, 116, 138-139
environmental sustainability 1-2, 4, 8, 10-13, 16, 20-21, 23, 75-78, 86, 112, 137, 139-140, 142-143, 145, 147, 149-153,

Index

156, 166, 172
environmentally conscious consumer behaviour 115
Environmentally Friendly Consumer Behavior 135
Eskom 125-126
E-Society 162-166, 173, 178, 183
E-Society infrastructure 162-164, 178, 183
eutrophication 86, 92, 94, 100-101, 107

F

fossil fuel-based 32

G

global warming 2, 4, 23, 36-37, 47, 64, 86, 92, 94-95, 100, 107, 113, 137-141, 143-147, 156
Globalization 1-20, 22-23, 93, 162-165, 172-173, 183
Green Consumer Behavior 112, 135
Green Consumerism 30
green consumers 114-115, 135
green gap 112-114, 116-119, 121-123, 125-127, 135
green lifestyle 123, 125-126
green products 113-116, 122-125, 127, 135
Green-Economy Production 151
greenhouse gases 32, 34, 36, 45, 95, 137-138, 142, 144, 150, 152, 156

H

hydropower 62, 66, 68-69, 71, 75, 77, 144, 149

I

India 17, 37, 40-41, 43, 46, 58-79, 84, 116, 126, 167, 169, 171
Indus 58-70, 72, 74-75, 77, 79, 84
Indus Waters Treaty (IWT) 61, 65-66, 69, 77, 79, 84
infrastructure 62, 73, 125, 141-142, 162-165, 167-168, 172-174, 178-179, 183
international business 13, 87
International Court of Arbitration (ICA) 69, 78, 84
international politics 60, 69, 86-87, 91
international relations 4, 12, 20, 33, 70, 86-87, 92-94, 107

J

Jhelum 59, 63-64, 67-69, 74, 77, 84

K

Kyoto agreement 41, 48
Kyoto Japan 34, 36

M

multilateral agreements 33-34

N

New Ecological Paradigm (NEP) 118
Non-Renewable Resources 156

P

Paris Accord 48-49, 51
Perceived Consumer Effectiveness (PCE) 118-119, 121-122, 135
Permanent Indus Commission (PIWC) 68, 77, 84
policies 8, 16, 18, 33-34, 38, 44, 51, 86, 88, 94, 100, 102, 107, 138, 142-143, 145, 147, 149-151, 153, 156, 165, 179
political ecology 1, 3-4, 7-8, 10, 12, 15, 21, 30
political problem 32
pollution 3, 11, 17, 19-20, 23, 35, 86-88, 91-97, 99-101, 107, 113, 115, 138,

259

140-144, 146-148, 150, 156
poverty 5, 21-22, 34, 93, 126, 163-166, 172, 174-179, 183
precipitation 86, 92, 94, 98-99
Pro-Environmental Consumer Behavior 135

R

Ravi 63-64, 74, 77, 84
Renewable Resources 156

S

sustainability 1-2, 4-5, 8, 10-13, 16, 19-23, 75-78, 86-88, 92-93, 97-99, 101, 103-105, 112-116, 123, 126-127, 135, 137-140, 142-143, 145, 147, 149-153, 156, 162-168, 171-172, 177-178, 183
sustainable consumption 1, 3, 10, 16-18, 21-23, 177
sustainable development 4, 22, 35, 88, 93, 112, 163-167, 169, 171-173, 175-179, 183

T

technology 4, 12-13, 15, 34, 41, 44, 47, 49, 102, 106, 163-165, 167, 173-174, 183
tension 60, 65, 69, 78, 84
The Permanent Indus Commission (PIWC) 68, 77, 84
Tsar Economic China 137, 147

U

UNFCCC 32, 34-37, 39, 45, 48
unsustainable environment 137, 156

V

virtual reality 164-165, 173

W

water scarcity 58, 62-64, 71, 75, 77-78, 84
water security 58, 64, 69, 71, 75, 77-78, 84
world's average temperature 32

Purchase Print, E-Book, or Print + E-Book

IGI Global books can now be purchased from three unique pricing formats:
Print Only, E-Book Only, or Print + E-Book. Shipping fees apply.

www.igi-global.com

Recommended Reference Books

Impact of Meat Consumption on Health and Environmental Sustainability
ISBN: 978-1-4666-9553-5
© 2016; 410 pp.
List Price: $210

Smart Cities as a Solution for Reducing Urban Waste and Pollution
ISBN: 978-1-5225-0302-6
© 2016; 362 pp.
List Price: $190

Geospatial Research
ISBN: 978-1-4666-9845-1
© 2016; 1,997 pp.
List Price: $1,845

Natural Resources Management
ISBN: 978-1-5225-0803-8
© 2017; 1,647 pp.
List Price: $1,850

Waste Management Techniques for Sustainability
ISBN: 978-1-4666-9723-2
© 2016; 438 pp.
List Price: $240

Climate Change Impact on Health and Environmental Sustainability
ISBN: 978-1-4666-8814-8
© 2016; 711 pp.
List Price: $325

Looking for free content, product updates, news, and special offers?
Join IGI Global's mailing list today and start enjoying exclusive perks sent only to IGI Global members.
Add your name to the list at www.igi-global.com/newsletters.

Publishing Information Science and Technology Research Since 1988

IGI Global
DISSEMINATOR OF KNOWLEDGE

www.igi-global.com   Sign up at www.igi-global.com/newsletters   facebook.com/igiglobal   twitter.com/igiglobal

Stay Current on the Latest Emerging Research Developments

Become an IGI Global Reviewer for Authored Book Projects

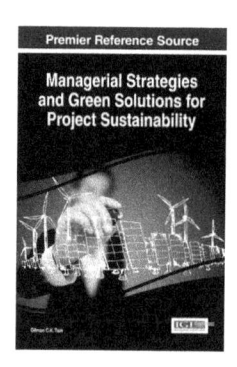

The overall success of an authored book project is dependent on quality and timely reviews.

In this competitive age of scholarly publishing, constructive and timely feedback significantly decreases the turnaround time of manuscripts from submission to acceptance, allowing the publication and discovery of progressive research at a much more expeditious rate. Several IGI Global authored book projects are currently seeking highly qualified experts in the field to fill vacancies on their respective editorial review boards:

Applications may be sent to:
development@igi-global.com

Applicants must have a doctorate (or an equivalent degree) as well as publishing and reviewing experience. Reviewers are asked to write reviews in a timely, collegial, and constructive manner. All reviewers will begin their role on an ad-hoc basis for a period of one year, and upon successful completion of this term can be considered for full editorial review board status, with the potential for a subsequent promotion to Associate Editor.

If you have a colleague that may be interested in this opportunity, we encourage you to share this information with them.

www.igi-global.com

InfoSci®-Books
A Database for Information Science and Technology Research

Maximize Your Library's Book Collection!

Invest in IGI Global's InfoSci®-Books database and gain access to hundreds of reference books at a fraction of their individual list price.

The InfoSci®-Books database offers unlimited simultaneous users the ability to precisely return search results through more than 80,000 full-text chapters from nearly 3,900 reference books in the following academic research areas:

Business & Management Information Science & Technology • Computer Science & Information Technology
Educational Science & Technology • Engineering Science & Technology • Environmental Science & Technology
Government Science & Technology • Library Information Science & Technology • Media & Communication Science & Technology
Medical, Healthcare & Life Science & Technology • Security & Forensic Science & Technology • Social Sciences & Online Behavior

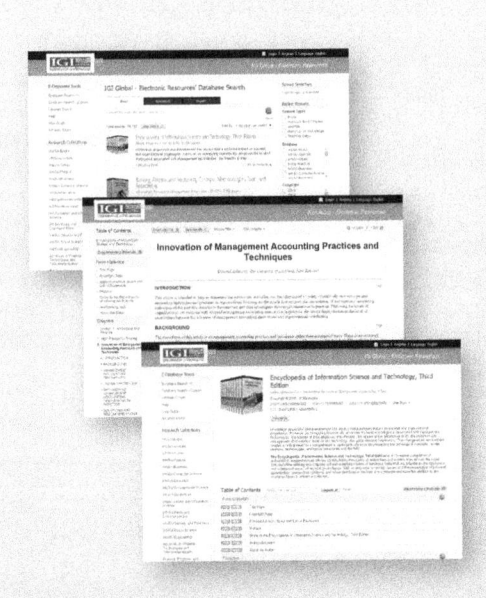

Peer-Reviewed Content:
- Cutting-edge research
- No embargoes
- Scholarly and professional
- Interdisciplinary

Award-Winning Platform:
- Unlimited simultaneous users
- Full-text in XML and PDF
- Advanced search engine
- No DRM

Librarian-Friendly:
- Free MARC records
- Discovery services
- COUNTER4/SUSHI compliant
- Training available

To find out more or request a free trial, visit:
www.igi-global.com/eresources

IGI Global
DISSEMINATOR of KNOWLEDGE
www.igi-global.com

IGI Global Proudly Partners with

eContent Pro International

www.igi-global.com

Enhance Your Manuscript with eContent Pro International's Professional Copy Editing Service

Expert Copy Editing

eContent Pro International copy editors, with over 70 years of combined experience, will provide complete and comprehensive care for your document by resolving all issues with spelling, punctuation, grammar, terminology, jargon, semantics, syntax, consistency, flow, and more. In addition, they will format your document to the style you specify (APA, Chicago, etc.). All edits will be performed using Microsoft Word's Track Changes feature, which allows for fast and simple review and management of edits.

Additional Services

eContent Pro International also offers fast and affordable proofreading to enhance the readability of your document, professional translation in over 100 languages, and market localization services to help businesses and organizations localize their content and grow into new markets around the globe.

IGI Global Authors Save 25% on eContent Pro International's Services!

Scan the QR Code to Receive Your 25% Discount

The 25% discount is applied directly to your eContent Pro International shopping cart when placing an order through IGI Global's referral link. Use the QR code to access this referral link. eContent Pro International has the right to end or modify any promotion at any time.

Email: customerservice@econtentpro.com

econtentpro.com

Information Resources Management Association

Advancing the Concepts & Practices of Information Resources Management in Modern Organizations

Become an IRMA Member

Members of the **Information Resources Management Association (IRMA)** understand the importance of community within their field of study. The Information Resources Management Association is an ideal venue through which professionals, students, and academicians can convene and share the latest industry innovations and scholarly research that is changing the field of information science and technology. Become a member today and enjoy the benefits of membership as well as the opportunity to collaborate and network with fellow experts in the field.

IRMA Membership Benefits:

- **One FREE Journal Subscription**
- **30% Off Additional Journal Subscriptions**
- **20% Off Book Purchases**
- Updates on the latest events and research on Information Resources Management through the IRMA-L listserv.
- Updates on new open access and downloadable content added to Research IRM.
- A copy of the Information Technology Management Newsletter twice a year.
- A certificate of membership.

IRMA Membership $195

Scan code or visit **irma-international.org** and begin by selecting your free journal subscription.

Membership is good for one full year.

www.irma-international.org